SpringerBriefs in Geography

SpringerBriefs in Geography presents concise summaries of cutting-edge research and practical applications across the fields of physical, environmental and human geography. It publishes compact refereed monographs under the editorial supervision of an international advisory board with the aim to publish 8 to 12 weeks after acceptance. Volumes are compact, 50 to 125 pages, with a clear focus. The series covers a range of content from professional to academic such as: timely reports of state-of-the art analytical techniques, bridges between new research results, snapshots of hot and/or emerging topics, elaborated thesis, literature reviews, and in-depth case studies.

The scope of the series spans the entire field of geography, with a view to significantly advance research. The character of the series is international and multidisciplinary and will include research areas such as: GIS/cartography, remote sensing, geographical education, geospatial analysis, techniques and modeling, landscape/regional and urban planning, economic geography, housing and the built environment, and quantitative geography. Volumes in this series may analyze past, present and/or future trends, as well as their determinants and consequences. Both solicited and unsolicited manuscripts are considered for publication in this series.

SpringerBriefs in Geography will be of interest to a wide range of individuals with interests in physical, environmental and human geography as well as for researchers from allied disciplines.

More information about this series at http://www.springer.com/series/10050

Sandipan Ghosh · Sanat Kumar Guchhait

Laterites of the Bengal Basin

Characterization, Geochronology
and Evolution

 Springer

Sandipan Ghosh
Department of Geography
Chandrapur College
Burdwan, West Bengal, India

Sanat Kumar Guchhait
Department of Geography
University of Burdwan
Burdwan, West Bengal, India

ISSN 2211-4165　　　　　ISSN 2211-4173　(electronic)
SpringerBriefs in Geography
ISBN 978-3-030-22936-8　　　ISBN 978-3-030-22937-5　(eBook)
https://doi.org/10.1007/978-3-030-22937-5

Cover image by Sonja Weber, München

This Springer imprint is published by the registered company Springer Nature Switzerland AG
The registered company address is: Gewerbestrasse 11, 6330 Cham, Switzerland

Foreword

It takes a brave man to write about laterite today. The material is very widespread in the world and is extensively used in buildings and roads, yet in many ways it remains a puzzle. There are arguments about what it is, and dictionaries and glossaries use definitions that are far from agreed. Laterite has been referred to as a soil type, a rock type, a complete weathering profile—such a range that it has led to calls for the term to be abandoned altogether. This won't happen because the material that looks like the Indian laterite occurs abundantly worldwide, and the term is so widespread among ordinary people, even if the meaning is unclear. It must be remembered that the first accounts of laterite, including that of Buchanan who coined the word, described a material, not a profile or a process.

The origins of laterite are permanently under debate, and there are numerous conflicting hypotheses. Perhaps several processes are involved that produce similar looking end products. The arguments have been going on for a very long time, almost since the first descriptions of laterite in India going back three hundred years. This is much longer than most scientific controversies.

Because of this conflict, occasional summaries are produced, which provide a basis for the next round of discussion, and this book is such a contribution. Some attempted summaries try to cover the whole world, and others are based on limited areas. The present book attempts to give a world overview at the beginning, but then describes the laterites of a specific area, the Bengal Basin, in detail.

Local studies are particularly valuable as they provide fine detail to illustrate the reality of the situation, and contrast with the often arm-waving generalizations of global debate. Each region where this is attempted has its own unique location, geology, and landscape history that can provide special opportunities, and also some limitations, on what conclusions can usefully be drawn.

In the study area presented in this book, we find a range of laterites, some formed on old bedrock and some on younger sediments of Cenozoic age, some containing fossils. To this, the authors have added results from optically stimulated luminescence (OSL) dating, which provides further constraints on age. These features make this area especially suited for studying the laterites of different ages, especially the younger ones.

Alas, the details presented here will not stop controversies, but the book presents a new step on the ladder leading to better understanding. They go all the way from definitions of terms, to detailed profile descriptions, presented in clear diagrams that provide a factual basis to inform our future arguments.

Armidale, Australia   Cliff Ollier
April 2019   Emeritus Professor
                                                         University of New England

Preface

The present work, entitled *Laterites of Bengal Basin: Characterization, Geochronology and Evolution*, is a comprehensive attempt to explore, analyse, discuss, and elucidate various dimensions of laterites and related ferruginous materials on the basis of detailed geomorphic, geological, and palaeogeographic characteristics (*viz.* lithology, structure, weathering profile, regolith geology, terrain morphology, Quaternary geomorphology, palaeoclimatology) of the Bengal Basin obtained from various sources such as maps, reports, books, research papers, as well as extensive field survey and profile analysis.

The studies were restricted in the western shelf zone of the Bengal Basin or western geomorphic unit of Ganga–Brahmaputra–Meghna Delta where the glimpses of Late Tertiary–Quaternary laterites cover the western districts of West Bengal (Birbhum, Paschim and Purba Bardhaman, Bankura, and Paschim Medinipur). The lateritic unit of this region is renowned as the *Rarh Plain* (*Rarh* means the land of red soil) which is dissected by the dense network of rills and gullies to develop badlands. The exhaustive field studies include reconnaissance traverses, detailed profiling or litho-log preparation of the important sections, characterization of individual sections and its relationship with the underlying and overlying lithologies, and age determination through dating techniques. More than twenty-seven sections of different laterites spread over the *Rarh Plain* had been studied and analysed to unearth the palaeogenesis of variable ferruginous facies in this part of the Bengal Basin.

Geomorphologically, the laterites are generally concentrated in the upland areas of dry deciduous forests or interfluves which are dissected by major west to east flowing peninsular rivers and their tributaries (*viz.* Dwarka, Mayurakshi, Kopai, Ajay, Kunur, Damodar, Dwarakeswar, Silai, Kangsabati, Subarnarekha rivers). The eastern part of the laterite occurrences in West Bengal shows a sharp contact with the Gangetic alluvium to the east. Small laterite hillocks on Rajmahal Basalt Traps represent butte-type structures. But all of the laterites are restricted within very low levels from 40 m to 100 m from mean sea level. The most interesting fact is that the laterites occur over a wide variety of rocks including basalts, granite–gneiss, Gondwana sandstones, dolerite dyke, and unconsolidated sediments of Late Tertiary and Early Quaternary age.

The laterites belong to two types: (1) in situ primary laterites and (2) *ex situ* secondary or reworked laterites. The primary laterites may be genetically related to four types of parent rocks, viz. (a) Rajmahal Basalt Trap, (b) dolerite in gneissic country, (c) Gondwana sequence, and (d) gneiss. The secondary laterites do not have direct genetic relation with the underlying rocks or sediments. The study reveals that the in situ lateritization on basalt is characterized by well-developed laterite profile (8 to 10 m thick) starting from top hardcrust/duricrust, mottled zone, lithomargic clay, and saprolite, followed by parent basalt at the bottom of these sections. The *ex situ* laterite developed over Tertiary or Quaternary sediments involved ferruginization of lower conglomerate–pebble horizon–sandstone unit but not the upper *Sijua* and *Panskura* alluvium sediments. The pisolitic hardcrust is appeared as primatry type in the laterite profile, but the vermicular hardcrust with tubular fluid passage channels are very poorly developed and absence of lithomargic clay or pallid zone is the feature of secondary type. So, for apposite understanding of lateritization and genesis of laterite lithosections in the western part of Bengal Basin, the present monograph has systematically presented the fundamental details, discussion, analysis, and conclusion in ten chapters.

In Chap. 1, an attempt has been made to present clearly the statement of research problem, its regional and global issues, major conceptual aspects of regolith geology, imperative ideas, and definitions of laterite and important terms related to laterite study. In Chap. 2 it tries to focus on the literature review, the identification of research gaps in the study of Indian laterites, and methodological outlook of the study. Next, in Chap. 3 it tries to unearth the tectono-geomorphic evolution of the Bengal Basin and its geological structure which are essential and fundamental part of the study. Additionally, the climate, soils, and natural vegetation of study area are discussed here. The effect of tropical weathering on the gneiss, basic dyke, Gondwana group of rocks, and Rajmahal Basalt Traps is minutely discussed in Chap. 4. Chapter 5 minutely has analysed the profiles of low-level or *ex situ* laterites which separate the lithological formations of Archaean, Gondwana, and Tertiary gravels from the Sijua and Chuchura Formations (Quaternary Alluvium). The various chemical properties of laterite samples, lateritization processes, and applicable theory of lateritization are discussed in Chap. 6 to get ideas about the genesis of ferruginous layers in the weathering profiles. In Chap. 7, the age determination, span of lateritization event, and dating data analysis are included to draw significant information about the geochronology of laterites. It tries to unearth the palaeogenesis, palaeoclimatic implication, and palaeogeomorphic evolution of laterites in the shelf zone of the Bengal Basin in Chap. 8. In Chap. 9, the economic significance of laterites, productivity of latosols, potentiality of geotourism, and soil erosion issue are discussed to get few inferences on the significance of laterite to human society. Chapter 10 is the epilogue part which summarizes and concludes the extensive field and laboratory studies of the important profile sections of western districts (viz. Birbhum, Paschim Bardhaman, Bankura, and West Medinipur) of West Bengal.

Burdwan, India

Sandipan Ghosh
Sanat Kumar Guchhait

Acknowledgements

The authors take the opportunity to express their deep sense of gratitude to Dr. Cliff Ollier (Emeritus Professor, University of New England, Australia) for providing his invaluable research papers and critical ideas about the regolith geology of tropics and the laterites of India. Professor Ollier has given new concepts, ideas, and research findings to understand the evolution of laterites in the Bengal Basin.

The authors like to extend their thanks to Prof. Dr. Mike Widdowson (Lecturer, Department of Geography and Geology, University of Hull, England) who has provided many valuable articles, chapters, and valuable ideas to carry out the fieldwork of laterite study.

The authors like to express their deep sense of gratitude to Dr. Suvendu Roy (Assistant Professor, Kalipada Ghosh Tarai Mahavidyalaya, Darjeeling), Subhankar Bera (Junior Research Fellow, Department of Geography, University of Kalyani), Rahaman Ashique Ilahi (Project Assistant, National Research Centre for Orchids, ICAR, Sikkim), Subhamay Ghosh (Research Scholar, CSRD, Jawaharlal Nehru University, New Delhi), Sukanta Mandal (Assistant Teacher), and Ankita Saha (Assiatnt Teacher) for their rigorous help, support, and cooperation in completing the fieldwork and data analysis.

The authors like to extend their thanks to all the organizations from which data, reports, and relevant maps have been collected, *viz.* National Bureau of Soil Survey and Land Use Planning (Kolkata), Geological Survey of India (Kolkata), Survey of India (Kolkata), National Atlas Thematic Mapping Organization (Kolkata), and National Remote Sensing Agency (Hyderabad).

The authors finally like to express their thanks and gratitude to all the faculty members of the Department of Geography, The University of Burdwan (Dr. Giyasuddin Siddique, Dr. Gopa Samanta, Dr. Narayan Chandra Jana, Dr. Biplab Biswas, Dr. Namita Chakma, Dr. Biswaranjan Mistri, Dr. Deb Prakash Pahari, Dr. Somasis Sengupta, Dr. Subodh Chandra Pal, Dr. Tapas Mistri, and

Dr. Sumana Sarkar) for their valuable suggestions from time to time, particularly for their constructive criticism and showing interest with regard to the present work and thought-provoking enquiry.

Sandipan Ghosh
Sanat Kumar Guchhait

Contents

Chapter 1
Introduction to Laterite Study

Abstract It is quite unfeasible to travel far in India without observing the remarkable ferruginous hardcrust to which Buchanan in 1807 gave the name of laterite. The reddish brown colour regolith (used as brick) with concentration of Fe–Al oxides has fascinated many researchers of earth sciences about its classification, evolution and variable occurrences on different geological formations. In Indian peninsula, laterite is a post-cretaceous stratigraphic succession with a polycyclic nature of evolution and it marks the unconformity with Late Quaternary alluvium. Before going into main discussion, it is necessary to understand the specific features and importance of laterite as a whole. The place of laterite in regolith science, definitions, ideas, profile and characterization of laterite and imperative terminology related to laterite study are discussed in this chapter.

Keywords Regolith · Laterite · Lateritization · Weathering · Ferricrete

1.1 General

Palaeoweathering studies contribute to palaeoenvironmental and palaeogeographical reconstructions, global correlation of deposits, records of global change, rate and timing of uplift and erosion, landscape evolution and an inventory of ore resources. The geological approach to ancient weathering features involves descriptions and study of complete geological formations, rather than just superficial levels, and frequently reveals weathering profiles that are much thicker and which have unusual geochemical signatures and palaeoclimatic proxies in comparison with present landscapes (Thiry et al. 1999). To analyze the above phenomena in recent decade, there is a revival of an inter-disciplinary approach rather than a subject, i.e. regolith geosciences or regolith geology, which principally deals with identifying potential zones of mineralization, determining mineral transport and transformation mechanistic processes in landforms, identifying palaeo-landscape features and process, groundwater and salinity issues and identifying natural geochemical hazards.

Of all the materials making up the upper part of solid earth, the ultimate product of weathering and the most important to humans are the surface mantle, the regolith

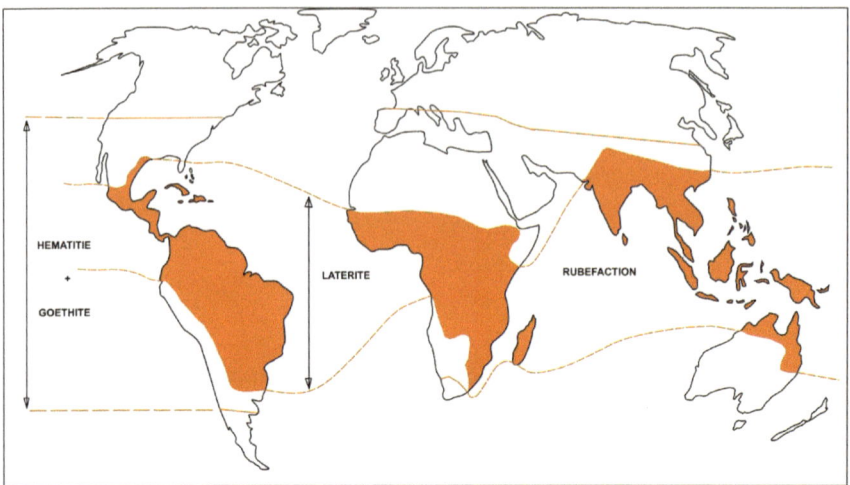

Fig. 1.1 The zone of rubefaction coincides with the intertropical area of laterite formation (modified from Tardy 1992)

(Taylor and Eggleton 2001; Scott and Pain 2009). The study of regolith has been used to interpret past geological and climatic conditions. So a new scientific approach has been developed, i.e. regolith geoscience or regolith geology, to unearth the genesis of weathering profiles and to understand the economic potentiality of the regolith. The study of regolith is able to provide insights to the evolution of the landforms and provides one of the few windows to our past which are so important in predicting how shifts in climate are likely to affect us (Taylor and Eggleton 2001).

The continents which were passing through the hot and humid tropical climatic zone in past must have the ferruginous materials or duricrust in the surface or sub-surface. Presently, the red and yellow ferralitic soils, tropical ferruginous soils, and other ferruginous formations such as ferricrete, bauxite, plinthite, carapace, mottled horizon, kaolinitic lithomarge, etc. are confined in the zone of rubefaction (Pedro 1968; Tardy 1992). The zone of rubefaction coincides with the intertropical area of laterite formation, characterized by the development of kaolinite and hydroxides, oxi-hydroxides or oxides of iron and aluminum (Fig. 1.1). Many scientists have used the term laterite to any ferruginous materials. The origin of the term 'laterite' was connected to the Malabar Coast of southern India where Buchanan (1807) found the hard brick like materials, named 'laterite'. But with the passage of time, the identification of laterite and further research on laterite give birth to numerous words and terms which should be explained here to understand the laterite profile minutely. This chapter deals with the definition, classification, the terminology of a related series of ferruginous materials, principal characterization, ideal weathering profile, and perennial problems in the investigation of laterites.

1.2 Regolith

The term 'regolith' was first defined by Merrill (1897). Extracts from his work read:

> Everywhere, with the exception of comparatively limited portions laid bare by ice or stream erosion, or on the steepest mountain slopes, the underlying rocks are covered by an incoherent mass of varying thickness composed of materials essentially the same as those which make up the rocks themselves…. In places this covering is made-up of material originating through rock weathering or plant growth in situ…. This entire mantle of unconsolidated material, whatever its nature or origin it is proposed to call the regolith….

Merrill also provided a tabulation of some regolith materials and their genesis (Table 1.1). There are two types of regolith—(1) sedentary regolith (residual deposits, cumulose deposits) and (2) transported regolith (colluvial deposits, alluvial deposits and Aeolian deposits).

Regolith is a general term for the layer of mantle of fragmental and unconsolidated rock material, whether residual or transported and of highly varies character, that nearly everywhere forms the surface of the land and overlies or covers the bedrock (Jackson 1997). It includes rock debris of all kinds, volcanic ash, glacial drift, alluvium, loess and Aeolian deposits, vegetal accumulations, and soil. Regolith is the entire unconsolidated or secondarily re-cemented cover that overlies more coherent bedrock and which has been formed by weathering, erosion, transport, and/or deposition of the older material (Taylor and Eggleton 2001; Scott and Pain 2009). In situ deep weathering is particularly common in the tropics, and there is a prevailing idea that tropics and deep weathering go together. Laterites or any ferruginous deposits of India are studied by regolith geology, and the deep weathered profiles of laterites generate many controversies related to genesis, characterization, and palaeogeographic inference. Weathering is an integral part of the geological cycle, and this book is concerned with the weathering, erosion, and deposition processes, that is regolith processes. Such processes involve the interaction between minerals, air, and

Table 1.1 Merrill's (1897) summary of his view of what he defined as regolith

The regolith	Sedentary	Residual deposits	Residuary gravels, sands, and clays, wacke, laterite, terra rosa, etc.
		Cumulose deposits	Peat, muck, and swamp soils, in part
	Transported	Colluvial deposits	Talus and cliff debris, materials of avalanches
		Alluvial deposits (including aqueo-glacial)	Modern alluvium, marsh, and swamp (paludal) deposits, the Champlain clays, loess, and adobe, in part
		Aeolian deposits Glacial deposits	Wind-blown material, sand dunes, adobe, and loess, in part Morainal material, drumlins, eskers, osars, etc.

Source Taylor and Eggleton (2001)

Fig. 1.2 **a** Weathering in the geological cycle and **b** the influence of different interactions on regolith (Scott and Pain 2009)

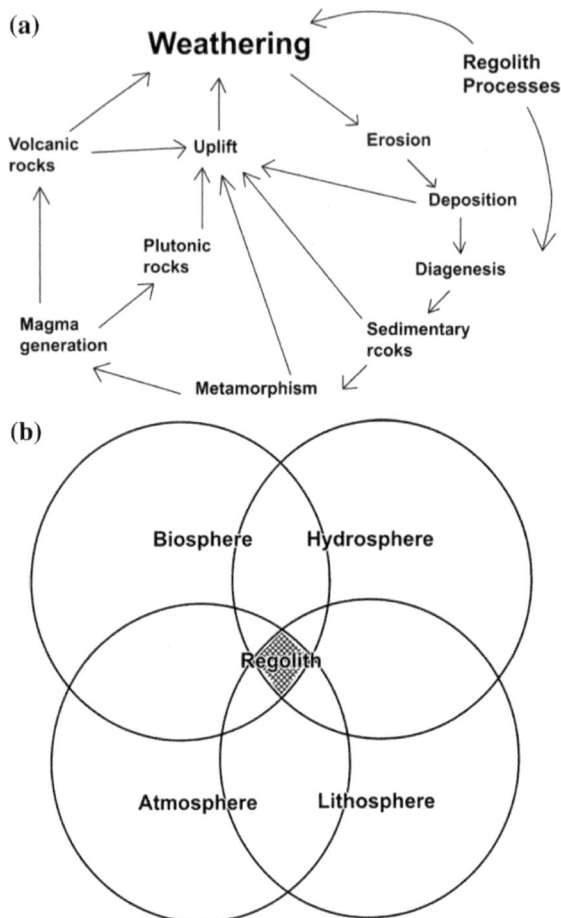

water, which is enhanced in most cases by the activities of biota (Fig. 1.2) (Scott and Pain 2009). Recent studies have shown that the regolith is much more complex than just a mantle of weathering products. It may be of allochthonous, autochthonous, or mixed allochthonous-autochthonous derivation. Generally, the greatest volume of the regolith is of allochthonous origin (Aleva 1994).

1.3 Laterite

The term 'laterite' was introduced by the British East India Company's surgeon Francis Buchanan in 1807 during a reconnaissance trip through the western part of Peninsular India (mainly in Angadipuram, Kerala). He found a soft reddish brown rock that could be cut with steel implements to produce bricks which after drying in the sun would harden irreversibly (Aleva 1994). Reviewing the different meanings that have been given to the term 'laterite' since Buchanan's original paper (Buchanan 1807), Schellmann (1982) proposed to give to this particular regolith the following definition: *'Laterites are products of intense subaerial weathering whose Fe and/or Al content is higher and Si content is lower than in merely kaolinized parent rock'*. Laterites consist predominantly of mineral assemblages of goethite, hematite, Al-hydroxides, kaolinite minerals, and quartz. Schellman's definition of laterites is perhaps most easily understood by considering the triangular diagram (Fig. 1.3) which is plotted the Si, Al, and Fe oxide contents of a particular unweathered rock and its weathering products.

The position in this diagram of the hypothetical 'merely kaolinized parent rock' of any laterite results from the following assumptions:

(1) The whole content in Al_2O_3 of the original unweathered rock is retained and is transformed into kaolinite;
(2) The whole content in Fe (calculated as Fe_2O_3) of the original rock is also preserved; and
(3) The Si content of the original rock, in excess of that necessary to combine with Al_2O_3 in kaolinite, is assumed to be leached out. However, if quartz is present in the rock, its Si content is also retained.

Ollier and Rajaguru (1989) noted that laterite can refer to both the indurated, mottled zone of saprolite, as well as to concretionary material in India, and to nodular,

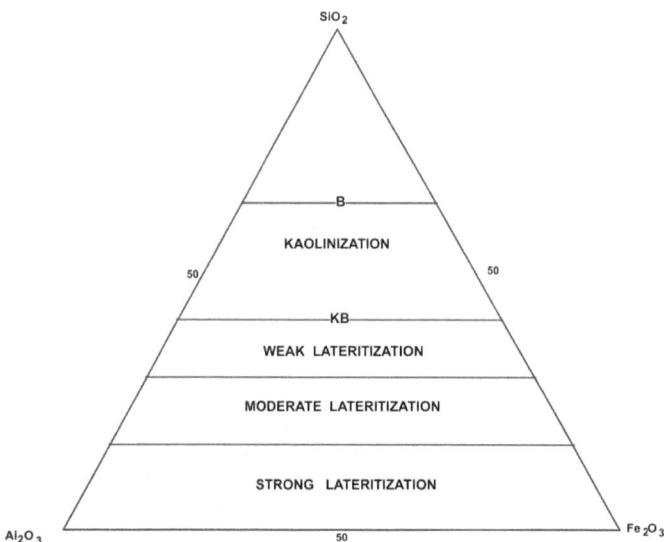

Fig. 1.3 Limits of kaolinization and lateritization in the case of basalt weathering (B—Basalt, KB—Kaolinized basalt (modified from Herbillon and Nahon 1988)

concretionary, and pisolitic material in Africa, along with indurated saprolite. It is considered to comprise highly weathered material enriched in secondary forms of iron, aluminium, or both, poor in humus, depleted of bases and combined silica, with or without non-diagnostic substances such as quartz, limited amounts of weatherable primary minerals or silicate cays and hard or subject to hardening on exposure (Sivarajasingham et al. 1962; Bourman 1993).

The various researches (Fermor 1911; Alexander and Cady 1962; McFarlane 1976; Aleva 1994) appear to have been too great an emphasis on defining laterite at the expense of detailed and systematic investigations of weathered and ferruginized materials, independently of perceptions coloured by the traditional views concerning their nature and conditions of formation. In India, the Director General of the Geological Survey of India proposed to the IGCP (International Geological Correlation—later Cooperation—Programme) Board to make a project entitled 'Lateritization Processes' in 1975, and the final report was published in 1986. It was found that by far the most common laterites of India were rich in Si, Al and Fe, resulting from the intense chemical weathering of the more common felsic to fairly mafic igneous, metamorphic, and sedimentary rocks. The term 'laterite' may ultimately disappear, to be replaced by ferricrete, ferruginous duricrust, and the so-called companion materials by weathered mantle, mottled zones, plinthite, and bleached zones (Bourman 1993).

1.3.1 *Important Ideas*

The preferred meaning of the term 'laterite' for various rocks is given in many works (Newbold 1844; Fermor 1911; Fox 1936; Alexander and Cady 1962; Bates and Jackson 1980), and it is necessary to state clearly the statements which provide few important ideas and definitions to understand uniqueness of laterites in India. Newbold (1844) provided a description of laterite, as studied by him during an official journey from Machilipatnam to Goa. The laterite of Beder (now Bidar) is a purplish or brick red, porous rock, passing into liver brown perforated by numerous sinuous and tortuous tubular cavities either empty or filled with grayish-white clay (Newbold 1844). The debris of laterites washed from the surface by rains is often seen accumulating in low situations and reconsolidating into nodular conglomerate. It is found that the alluvial laterite is seen in all low-altitude areas of southern India and is easily to be distinguished by its nodular and pisiform character, its position and the thinness of its beds from the true laterite, as the reconsolidated debris of quartz, mica, and feldspar is from the true granite rocks, to accumulate in beds of some thickness tenacity (Newbold 1844).

The important ideas of Fermor (1911), Fox (1932, 1936), Alexander and Cady (1962), Sivarajasingham et al. (1962), Maignien (1966), Persons (1970), Gidigasu (1976), McFarlane (1976) and Aleva (1994) are summarized here on laterites to clarify much confusion about its terminology. These ideas are necessary to mention below:

- The term 'laterite' is used in two ways, namely stratigraphically as the name of a geological formation or unit and petrographically as the name of a tropical superficial weathered rock, i.e. regolith. Laterite can be said to be an end product or perfect regolith or extreme example of weathering processes taking place over large part of the earth's surface having tropical wet-dry climate.
- Laterite is formed by a geochemical process, and certain rocks undergo superficial decomposition under tropical humid climate, with the removal in solution of combined silica, lime, magnesia, soda, potash and with the residual accumulation, by capillary action, metasomatic replacement, and segregative changes of a hydrated mixture of Fe, Al, Mn—oxides and hydroxides.
- There is often a gradation in composition between true laterite as defined above and lithomarge, which is taken as the amorphous compound of composition $2H_2O \cdot Al_2O_3 \cdot 2SiO_2$, corresponding to the crystalline mineral kaolinite of the same composition.
- The property of hardening on exposure to the air is characteristic of many varieties of laterite, but it is not an essential property, from some laterites do not exhibit it.
- Laterites formed by the accumulation of detritus from masses of chemically formed laterite either alone or mixed with extraneous material, such as fragments of quartz or gneiss, may be termed detrital laterite.
- High-level ferruginous crusts occur as cappings on high ridges and on peneplain fragments. Foot-slope or colluvial, seepage-cemented ferruginous crusts, formed by cementation of colluvial materials that commonly contained fragments of crust, broken from a peneplain crust of a higher level.
- A distinction should be made between the laterite and saprolite to characterize the weathering of rocks into regolith. Few scholars lists the chemical—mineralogical reactions—(1) kaolinization of Al–Si-bearing minerals, (2) formation of Fe oxides from Fe-containing minerals, (3) formation of Al hydroxides by incongruent solution of kaolinite minerals, (4) congruent dissolution of kaolinite minerals, and (5) dissolution of quartz. The reactions 1, 2, 3, and 5 cause an enrichment of iron and aluminum in the saprolite, and reaction 4 increases the iron content in laterite.
- Now in broadest sense, the term 'laterite' includes ferricrete, iron—aluminum duricrusts, mottled horizon, carapace, curiasse, plinthites, pisolite, or nodule-bearing materials and also kaolinitic lithomarge.

1.3.2 Characterization

Laterites are the final product of sub-aerial alternation or weathering under tropical wet-dry climate. After their formation, many laterites have been covered by younger sediments, or they have been further transformed under subsequent different climatic conditions. Table 1.2 provides an overview of the position of laterite with the regolith, focusing on an overview of its contrasting components and complex stratigraphy. The regolith itself is defined as the upper, unconsolidated part of the earth between the consolidated rocky core and the atmosphere, including recent to sub-recent soil.

Table 1.3 indicates the variability of almost all the characteristics that comprise a laterite. This large variability is possibly the most characteristic feature of a laterite. Table 1.4 indicates a process-oriented characterization of laterite and its key features.

Laterite is the reddish brown coloured product of intense tropical weathering made up of mineral assemblages that may include Fe or Al oxides, oxyhydroxides or hydroxides, kaolinite, and quartz, characterized by a ratio $SiO_2 : R_2O_3$ (where R_2O_3 = $Al_2O_3 + Fe_2O_3$) and subject to hardening up on exposure to alternate wetting and drying (Ghosh and Guchhait 2015). Mineralogically, the in situ laterite of study area is essentially a mixture of varying proportions of goethite [FeO(OH)], haematite (Fe_2O_3), gibbsite (Al_2O_3, $3H_2O$), boehmite [AlO(OH)], limonite [γ-FeO(OH)], and kaolinite [$Al_2Si_2O_5(OH)_4$]. It is suggested that weathered materials having $Fe_2O_3 :$ Al_2O_3 ratio more than 1 and $SiO_2 : Fe_2O_3$ ratio less than 1.33 be termed as 'ferruginous laterites', while those with $Fe_2O_3 : Al_2O_3$ ratio less than 1 and $SiO_2 : Fe_2O_3$ less than 1.33 as 'aluminous laterite' (Karunakaran and Sinha Roy 1981; Ghosh and Guchhait 2015). It has been observed that the vermicular laterites of study area have important ferruginous materials, e.g. gibbsite, haematite, goethite, and limonite, having high percentage of Al_2O_3 and Fe_2O_3 and high percentage of kaolinite as the base of profile (Table 1.5).

1.3.3 Classification and Nomenclature

The IGCP Project 129 provided an opportunity to discuss the classification of laterites—as defined by Schellmann—at a lower, more detailed level. That concept was accepted by Herbillon and Nahon (1988), and they maintained that it accommodates

Table 1.2 The place of laterite within the regolith

Within the regolith four broad rock units may be distinguished on the basis of their formation history	
Unit I	Stratigraphically, the lowermost one is the unweathered bedrock, which can be of almost any age. It may outcrop in the present-day landscape, but generally, it is covered by a succession of younger formations, which are either autochthonous or allochthonous. The overlying units, number II to IV, comprise the regolith
Unit II	It represents a fossil, autochthonous, or residual weathering mantle with a succession of zones, resulting from former episodes of weathering. Where undisturbed, the upper zone of this unit is the laterite/bauxite formation
Unit III	It is transitional nature and of intermediate age. The lower subunit includes a downslope movement of the unconsolidated materials, having stone line. The upper subunit incudes the main mass of transported matter, having loose aggregate of detritus and variable composition and age
Unit IV	It is the recent to sub-recent reaction zone of the underlying regolith units, where they are in contact with today's atmosphere: the soils

Source Aleva (1994)

Table 1.3 Macroscopic characteristics of laterites for field use

Laterites are formed near the earth surface; they mostly occur just below the ferruginous soil. After their formation, they are either denuded by erosion or covered below younger deposits. They may be covered with a forest, with the tree roots deeply penetrating into and through the laterite

Hardness	It is highly variable, both within and between laterite deposits; from plastic, brittle, sectile, and breakable between the fingers to difficult to break with a hammer
Colours	These are highly variable, although mostly reddish, reddish brown, brownish to yellow brown in hue
Grain size	It is often difficult to assess, e.g. where laterite forms a massive, hard, or micro-crystalline mass, or where it is plastic or doughy. Grain size varies between <0.1 and 2 mm
Fabric	It is highly variable, from massive to even-grained and layered, but also with vermiform, scoriaceous, columnar, and root-like structures; if spherical bodies (pisoliths or preferably locally pisoids) occur, they may represent between 1 and 90% of the volume of the laterite
Chemical	The composition is highly variable, with Fe_2O_3 content between 1 and 60% and Al_2O_3 content between >60% (bauxite) and <10%
Mineralogical	They are mainly composed of newly formed minerals, such as gibbsite, goethite, hematite, maghemite, kaolinite, and secondary quartz
Loss if ignition (bound water)	It is variable with up to 34% in pure bauxite to circa 25% in goethitic and kaolinitic laterite
Unweathered rock-forming silicates	Feldspars, feldspathoids, hornblende, pyroxcne, biotite, garnet, etc. may occur as relict minerals of the original parent rock

Source Aleva (1994)

the concepts of 'relative' (accumulation of Fe_2O_3 and Al_2O_3 with loss of SiO_2 and bases) and 'absolute' (imports of Fe_2O_3 and/or Al_2O_3) accumulations (Bourman 1993). There was general consensus that such classification should be related to the type of parent rock: through its variations in mineral and chemical composition and its fabric, the parent rock impresses a distinct signature on the resultant laterite, e.g. a nickel laterite or bauxite (aluminium laterite). The nature of weathering processes, i.e. dissolution, may play havoc, however, with the concept of full autochthony, which is related to in situ weathering. There is the transformation of feldspar, pyroxene, and hornblende crystals that is replaced in situ by gibbsite, i.e. pseudomorphism (Aleva 1994).

This discussion on classification may seem theoretical, but in practice, the term 'laterite' is used frequently in an erroneous way, in particular for ferruginous deposits at or near the surface of planation surfaces or at the bottom of local slopes. Such iron-

Table 1.4 A process-oriented characterization of laterite

Laterites may be considered as metasomatic rocks, formed at or near the land surface, for which the essential process parameters are estimated to be:

- High average ambient temperature, in the order of 25–40 °C and sufficient organisms and decaying organic matter to make the abundant percolating rainwater a chemically and physically active fluid
- Active metasomatism—the process of practically simultaneously capillary solution and deposition by which a new mineral of partly of wholly different chemical composition may grow in the body of an old mineral or mineral aggregate
- High supply rate of rain, i.e. the solvent, (e.g. >1200 mm/year rain during at least, 9 months per year), to promote sufficient leaching activity
- A surface relief with a minimum of slope to minimize erosional incision resulting from the abundant precipitation
- High rate of percolation of the water in order to evacuate the leachate; favourable fabric allowing a continuous and pervasive flow of water
- Sufficient topographic height above the local and/or regional base level of erosion to promote a continuous high rate of percolation of water and transport of the leachate
- Containing sufficient soluble Al and Fe present in silicates, to form a skeleton of newly formed hydroxides and oxides of Al and Fe (and Ni and Mn)

Source Aleva (1994)

Table 1.5 Proportion of chemical components (in per cent) in the horizons of laterite profile on Gneiss at Pansiuri, Birbhum district

Sample	SiO_2	Al_2O_3	Fe_2O_3	FeO	CaO	MgO	TiO_2
Laterite at 2 m depth	10.23	39.86	32.05	0.46	0.28	0.32	2.13
Kaolin at 11 m depth	46.67	35.82	1.43	0.22	0.53	0.26	1.18
Bedrock-Gneiss at 31 m depth	61.92	20.10	3.83	1.62	0.78	3.25	0.65

Source Ghosh and Guchhait (2015)

rich, hard crusts are not the product of in situ weathering, but are formed from iron-rich solutions and debris derived from upslope former laterites. Ideally, it should be possible to classify lateritic materials in the field and not be dependent upon detailed and expensive laboratory analyses. It is the researcher's belief that an individual ferricrete reflects both the character of the original material that has been ferruginized and the accumulated effects of subaerial processes.

Laterite versus laterite is a peculiar distinction made by De Swardt (1964) who worked in the African Shield. Two main laterites are classified in Africa—(1) the older is a primary deposit (high-level laterite) which now forms cappings on erosion remnants and is now separated from the lower laterite scarp and (2) the younger (low level laterite) is composed largely of reworked material from the older formation which has been recemented to a hard pavement. McFarlane (1976) distinguished three types of laterites: continuous phase, pedogenetic, and groundwater laterites.

(1) Continuous phase laterites are composed of a coherent skeleton of the indurated elements. Laterites are characterized by a number of non-directional structures, e.g. vesicular, cellular, vermiform, tubular, etc.

Table 1.6 Laterite classification scheme (Bourman 1993)

I. Simple ferricrete	1. Ferricreted bedrock
	2. Ferricreted sediment
	a. Ferruginized clastic quartzose sediment including blocks of reworked ferricrete (detrital ferricrete)
	b. Ferruginized organic sediment displaying a massive to vesicular fabric (bog iron ores)
	c. Ferricrete formed by in situ weathering of iron-rich sediment containing siderite, glauconite, and pyrite
II. Complex and composite ferricrete	1. Pisolitic ferricrete in which pisoliths are important constituents
	2. Nodular ferricrete
	3.. Slabby ferricrete
	4. Vermiform ferricrete

Table 1.7 Laterite classification scheme (Pullan 1967)

I. Primary lateritic ironstone	Developed in situ from different parent materials with or without the introduction of extremely derived iron: vermicular; vesicular; cellular; cellular nodular; nodular; oolitic; and pisolitic
II. Secondary lateritic ironstone	Developed from the partial destruction of primary lateritic ironstone, transportation of fragments of the original ironstone, their subsequent deposition, and recementation with iron compounds: recemented conglomeratic; recemented breccia; and platy

(2) Pedogenetic pisoliths are considered to be formed within soil, which originally overlaid the underlying saprolite, a location where alternating conditions of wetting and drying would occur.

(3) Groundwater pisoliths would be formed in the zone of fluctuation of the groundwater table, in both the soil (as a closely packed layer directly overlying the saprolite) and the underlying saprolite (as more or less widely spaced pisoliths).

Bourman (1993) classified ferricrete into simple (ferruginized bedrock and ferruginized clastic and organic sediments—vesicular ferricrete) and complex (pisolitic, nodular, slabby, and vermiform) ferricrete (Table 1.6). Mottled and pallid zones in bedrock and sediment are treated as independent from the ferricrete. The morphological classification scheme of Pullan (1967) recognized vermicular, vesicular, cellular, cellular nodular, oolitic, and pisolitic (primary forms), recemented nodular, recemented conglomeratic, recemented breccias, and platy (secondary forms) lateritic ironstone as well as ferruginized rock and sediment (Table 1.7).

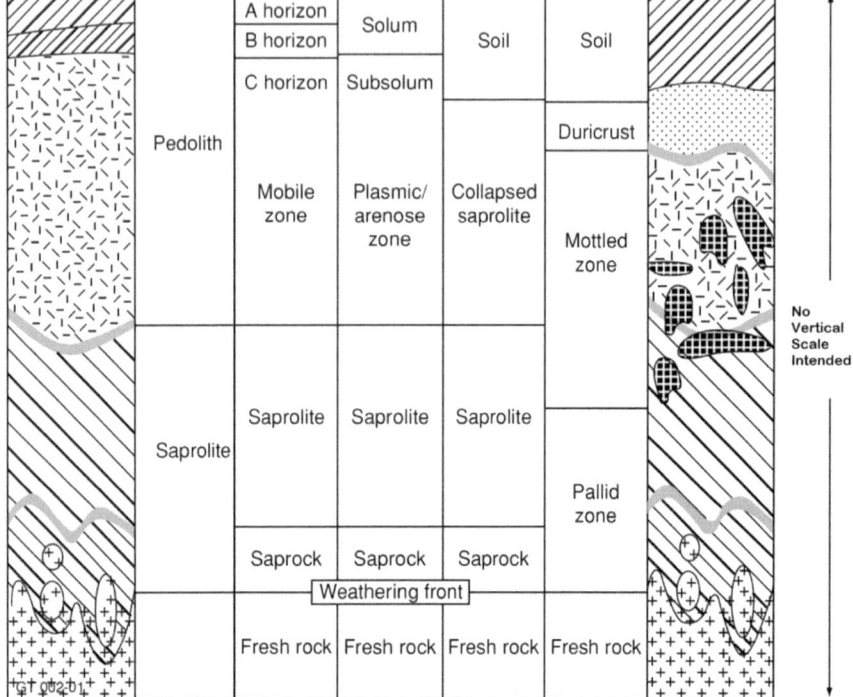

Fig. 1.4 A schematic diagram showing the variety of terms used to refer to parts of a tropical regolith profile, i.e. similar to laterite profile (modified from Taylor and Eggleton 2001; Taylor 2011)

1.4 Important Terminology in Laterite Study

The term 'laterite' was first published in scientific literature in nearly 210 years ago. There are now very extensive scientific literatures to understand the variable aspects of laterite and lateritization processes. The laterite literature is heavily encumbered by problems of terminology, and often, different materials are described by a single term to avoid overloading the large laterite vocabulary. Now, the earth scientists prefer to use the term 'laterite profile' instead of only 'laterite' to study all the ferruginous weathered profiles and horizons, because laterite profile includes the sections of weathered rocks, pallid zone, mottle zone, and ferricrete (i.e. duricrust). So before going into main section, it is utmost necessary to understand and to clarify the usable terms in the regolith geology of laterite profile (Fig. 1.4). These terms and their descriptions are mainly collected from the popular works by Alexander and Cady (1962), Paton and Williams (1972), McFarlane (1976), Tardy (1992), Aleva (1994), Eggleton (2001), Taylor and Eggleton (2001), and Gutierrez (2005).

- **Carapace**—It is a French term referring to the lower, poorly indurated horizon of the ferruginous zone of a laterite profile. It consists of ferruginous nodules and pisoliths in a weakly cemented matrix of kaolinite and iron oxides and oxyhydroxides.

- **Concretion**—A hand, compact, rounded, normally sub-spherical (but may be any shape) mass or aggregate of mineral matter generally formed by precipitation from aqueous solution (commonly about a nucleus or centre) within a rock or regolith and generally of a composition widely different from that of the rock in which it is found and from which it is rather sharply separated.

- **Core-stone**—It is a rounded, ellipsoidal, or broadly rectangular block, composed of virtually fresh parent rock in sparock or saprolite; the residual relatively unweathered remnant of a joint block, originating from any massive type of parent rock, e.g. granite, basalt, etc., but separated from it.

- **Cuirasse**—It is the highly indurated upper facies or horizon of the ferruginous zone of a lateritic regolith, with a massive pisolitic, nodular, or vesicular fabric.

- **Detritus**—It is the collective term for loose rock and regolith material that is worn off or removed by mechanical means, as by disintegration or abrasion, especially fragmental material, such as gravel, sand, silt and clay, derived from older rocks, or regolith and moved from their place of origin.

- **Duricrust**—Duricrust is a term coined by Woolnough (1930) to define a product of terrestrial processes within the zone of weathering in which either Fe and Al sesquioxides (in the case of ferricrete or bauxite) or silica (in the case of silcrete) or calcium carbonate (in the case of calcrete) or like above have dominantly accumulated in and/or replaced a pre-existing soil, rock, or weathered material, to give a substance which may ultimately develop into an indurated mass. Cellular duricrust is characterized by irregular to rounded bladder, cell, or bubble-shaped voids. It may contain pisoliths and/or nodules and show development of a mottled fabric. Fragmental duricrust has a fragmental or blocky fabric in outcrop and/or hand specimen. The interstices between fragments are commonly occupied by a clayey, ferruginous, or sandy matrix. Massive duricrusts have a homogenous fabric at the hand-specimen scale. It usually contains minor amounts of vesicles and tubules which may be filled or partly filled with clay and/or other sediment.

- **Ferricrete**—Ferricrete is the term to describe the surface or near-surface masses of regolith cemented by Fe-oxides and oxyhydroxides. Ferricrete is a nodular Fe-rich accumulation in which the Fe-nodules are indurated or cemented concretions consisting mainly of haematite in the porosity between kaolin crystals. Pisolitic ferricretes are cemented accumulations of pisoliths which are formed as clastic components from the materials where they formed by deposition of lamina of various compositions (goethite, kaolin, gibbsite, haematite, and magnetite with quartz grains). Nodular ferricrete contains nodules of earthy appearance that are smaller than pisoliths. Vermiform ferricrete consists of vermicular tubles filled or lined with kaolin and gibbsite set in a surface lag of pisoliths, bedrock, ferruginous clasts, and hardened mottles cemented by high Al-goethite.

- **Glaebule**—It is a three-dimensional compound unit within the matrix of a soil material or occurring as a discrete physical fabric element, generally approximately equant or prolate in shape and with a sharp boundary.
- **Ironstone**—It is highly ferruginous weathered material consisting mainly of iron oxides and oxyhydroxides, with variable amounts of aluminium hydroxides, silica, and phyllosilicates. Example is a part of a laterite profile, essentially conformable with the land surface, i.e. lateritic ironstone or duricrust.
- **Laterite Profile**—Laterite profiles are a vertical sequence of regolith facies showing some or all of the following, from the bottom-up: bedrock, saprock, saprolite, plasmic zone, mottled zone or ferruginous saprolite, lateritic residuum (Fig. 1.4). The successive zones of the laterite profile are separated by more or less clearly discernable interfaces. These interfaces appear to be reaction fronts in the sense of the 'weathering front'. In typical lateritic profiles have distinguished three zones—(a) zone of alteration (coarse saprolite, lithomarge) at the base, (b) a glaebular zone (ferricrete, mottled zone, lithorelicts) , located in the middle part, and (c) a soft zone (relative accumulation of primary minerals or secondary minerals; degradation and dismantling of glaebular material), non-indurated, located in the higher part of the profile.
- **Lateritic Residuum**—It is a collective term for the ferruginous part of a laterite profile, composed dominantly of oxides and oxyhydroxides of Fe or Al (goethite, hematite, maghemite, gibbsite, boehmite) with or without quartz. The term includes both fragments and duricrust developed essentially by residual processes and therefore has broad genetic and/or compositional relationship with the substrate.
- **Latosol**—It is a zonal soil characterized by deep weathering and abundant hydrous oxide material, developed under forested humid tropical conditions.
- **Lithomarge**—It is the compact, massive, generally kaolinitic clay. It has been applied in the French and Indian literature to clay-rich zones of the regolith, particularly in the upper saprolite. Lithorelict is an unweathered fragment of rock in an assemblage of secondary minerals.
- **Liesegang Rings**—It is nested rings or bands of yellow/brown/red colour in weathered rocks, generally caused by the precipitation of iron oxides and oxyhydroxides from solution.
- **Mottles**—Mottles are weakly cemented accumulations of minerals coloured differently from the body of the regolith. Mottles are the segregation of sub-dominant colour different from the surrounding region's colour. In regolith, mottles may have sharp, distinct, or diffuse boundaries. They are typically range in size from 10 to 100 mm, but may reach several metres in size. Mottled zone underlies the lateritic residuum and is commonly above the plasmic zone. They most commonly form by the precipitation of Fe^{3+} or less commonly Mn^{3+} in the regolith plasma (fine-grained phases like clay minerals).
- **Nodular**—Nodule is a lump of regolith, generally pebble-sized, not rock, different from its immediate surrounds. Nodular is composed of nodules, consisting of scattered to loosely packed nodules in a matrix of like or unlike charter, characterized by lumps, flocculated material, roundish aggregations, or large coated grains, often composed of the same material that encloses them, e.g. nodular ferricrete. Nodules are concretions with no regolith in which they form.

- **Pallid Zone**—Pallid means very pale to white, lacking pigmentation. Pallid zone is a zone or portion of a weathering profile that lacks colour. Its position in the profile and its mineralogy may vary, but the zone is generally dominated by kaolin and quartz. In a laterite profile, the pale-coloured region below the mottled zone, incorporating parts of the plasmic zone and/or sparolite has been referred to as the pallid zone.
- **Pisolith**—It is a spherical or ellipsoidal body resembling a pea in shape and limited in size to between 2 mm and about 64 mm in diameter. It may have a concentric internal structure, but concentric lamination is not diagnostic; however, most pisoliths have an outer cortex or skin.
- **Plasmic Zone**—It is mesoscopically homogenous part if a weathering profile, having clay (plasma) as a significant components, which has neither the lithic fabric of the saprolite nor the significant development of secondary entities such as nodules or pisoliths.
- **Plinthite**—It is iron-rich, humus-poor mixture of clay and quartz, commonly occurring as mottles, firm but uncemented when moist but hardening irreversibly into an ironstone hardpan or irregular aggregates on repeated wetting and drying.
- **Saprock**—Saprock is the first stage of weathering. It consists of partially weathered minerals and as yet unweathered minerals. Saprock maintains all the fabric and structural features of the rock.
- **Saprolite**—Saprolite is weathered bedrock in which the fabric at the parent rock, originally expressed by the arrangement of the primary mineral constituents of the rock, is retained. Compared to saprock, saprolite has more than 20% of weatherable minerals altered and generally collapses under a light blow. The presence of saprolite implies that weathering has been essentially isovolumetric. Here, weatherable minerals are wholly or partially pseudomorphed by clays and/or oxides and oxyhydroxides so that rock fabric is maintained.
- **Stone line**—It is layer in the regolith composed of gravel-size angular to subrounded fragments of weathering-resistant rock, commonly quartz and normally occurring at a depth between 0.3 m and several metres below a gently sloping ground surface. Stone line has been interpreted as the boundary between in situ weathered parent rock (saprolite) below an originally residual soil layer gradually moving downslope.
- **Vermiform**—It has the form of a worm. In the regolith, a fabric consists of tubes, pipes, or worm-shaped voids which may be filled or partly filled with, e.g. clays, sandy sediments, or iron oxides.
- **Weathering Profile and Front**—Rocks weather when they interact with air, water, and biota in the near-surface environment, having decline interacting with depth. There is a progressive change in weathering from surface to interior, and the physical extent of that region of change is called weathering profile. The weathering front is the boundary between fresh and weathered rock. The weathering of homogenous rocks like granite and basalts proceeds down joints in the rock, so the weathering front is usually irregular and deeply indented. But in such rocks, it is common for lumps of fresh rock with spherical or ellipsoidal shapes (corestones) to occur detached from the fresh rock below the front.

References

Aleva GJJ (1994) Laterites—concepts, geology, morphology and chemistry. ISRIC, Wageningen

Alexander LT, Cady JG (1962) Genesis and hardening of laterite in soils. Tech Bull (United States Department of Agriculture) 1281:1–90

Bates RL, Jackson JA (1980) Glossary of geology. American Geological Institute, Virginia

Bourman RP (1993) Perennial problems in the study of laterite: a review. Aust J Earth Sci 40(4):387–401

Buchanan F (1807) A journey from Madras through the countries of Mysore, Kanara and Malabar (3 volumes). East India Company, London

de Swardt AMJ (1964) Lateritization and landscape development in parts of equatorial Africa. Zeitschrift fur Geomorphologie 8:313–333

Eggleton RA (2001) The regolith glossary. CRC Australia CRC LEME, Canberra

Fermor LL (1911) What is laterite? Geological Magazine 8:454–462

Fox CS (1932) Bauxite and aluminous laterite. Crosby Lockwood & Son, London

Fox CS (1936) Buchanan's laterite of Malabar and Kanara. Records of the Geological Survey of India 69:389–422

Ghosh S, Guchhait S (2015) Characterization and evolution of primary and secondary laterites in northwestern Bengal Basin, West Bengal, India. J Palaeogeogr 4(2):203–230

Gidigasu MD (1976) Laterite soil engineering: pedogenesis and engineering principles. Elsevier, Amsterdam

Gutierrez M (2005) Climatic geomorphology. Elsevier, Amsterdam

Herbillon AJ, Nahon D (1988) Laterites and lateritization processes. In: Stucki JW, Goodman BA, Schwertmann U (eds.) Iron in soils and clay minerals. Springer, Dordrecht, pp 779–796

Jackson JA (1997) Glossary of geology. American Geology Institute, Alexandria

Karunakaran C, Sinha Roy C (1981) Laterite profile development linked with polycyclic geomorphic surfaces in South Kerala. In: Proceedings of the international seminar on lateritisation processes, Trivandum, India

Maignien R (1966) Review of research on laterites. UNESCO, Paris

McFarlane MJ (1976) Laterite and landscape. Academic Press, London

Merrill GP (1897) Rocks: rock-weathering and soil. MacMillan Company, New York

Newbold TJ (1844) Summary of the geology of Southern India. J R Asiat Soc 8:138–171

Ollier CD, Rajaguru SN (1989) Laterite of Kerala (India). Geogr Fis Dinam Quat 12:27–33

Paton TR, Williams MAJ (1972) The concept of laterite. Ann Assoc Am Geogr 62(1):42–56

Pedro G (1968) Distribution des principaux types d'alteration chimique a la surface du globe. Presentation d'une esquisse geographique. Rev Geogr Phys Geol Dyn 2(10):457–470

Persons BS (1970) Laterite—genesis, location, use. Plenum Press, New York

Pullan RA (1967) A morphological classification of lateritic ironstones and ferruginized rocks in Northern Nigeria. Niger J Sci 1:161–174

Schellmann W (1982) Eine neue laterite definition. Geol jahrb D 58:31–41

Scott KM, Pain CF (2009) Regolith science. Springer, Dordrecht

Sivarajasingham S, Alexander LT, Cady JG, Cline MG (1962) Laterite. Agronomy 14:1–56

Tardy Y (1992) Diversity and terminology of laterite profile. In: Martini IP, Chesworth W (eds) Weathering, soils and paleosols. Elsevier, Amsterdam, pp 379–405

Taylor G, Eggleton RA (2001) Regolith geology and geomorphology. Wiley, Chichester

Taylor G, Eggleton RA (2011) Regolith geology and geomorphology. Wiley, New York

Thiry M, Schmitt JM, Simon-Coincon R (1999) Problems, progress and future research concerning palaeoweathering and palaeosurfaces. In: Thiry M, Schmitt JM, Simon-Coincon R (eds) Palaeoweathering, palaeosurfaces and related continental deposits. Blackwell Science, Oxford, pp 3–17

Woolnough WG (1930) The influence of climate and topography in the formation and distribution of products of weathering. Geol Mag 67:123–132

Chapter 2
Literature Review and Research Methodology

Abstract The reddish brown colour regolith (used as brick) with concentration of Fe–Al oxides has fascinated many researchers of earth sciences about its classification, evolution, and variable occurrences on different geological formations. It is quite impossible to travel far in India without observing the remarkable ferruginous crust ferruginous crust, ferruginous gravels and red soils. Throughout the field examination, one significant question is raised about the nomenclature and genesis of laterites in the study area, as well as in West Bengal—how that ferruginous formation was formed in geological past. In this section, it is tried to focus on the literature review, the identification of research gaps in study of Indian laterites, and methodological outlook of the present study.

Keywords Laterite · Literature review · Research gap · Research methodology · *Rarh Bengal*

2.1 Literature Review

The study regarding laterites is considered as the most debatable and conspicuous matter in research due to its variable occurrences and types on various parent materials. A number of significant studies have been carried out by various geoscientists of the country and abroad to study laterites, analysing and comparing different aspects of ferruginous materials and laterite genesis. The research area of regolith geology includes the consideration of identification and mapping of laterites, factors and processes of lateritization, understanding the unique character of laterite, morphology of laterite profile, dating of ferruginous facies, genesis and geomorphic evolution of laterites and the palaeogeographic importance of laterites in India.

2.1.1 International Study

The term laterite has been applied to such a diverse array of geomorphic features that it no longer has value as a precise descriptive term (Paton and Williams 1972). The term 'laterite' was originally suggested by Buchanan (1807) as a name for highly ferruginous deposits first observed in Malabar in India. Maignien (1966) wrote a book, named '*Review of Research on Laterites*', mentioning different definition of laterites, scope of the problem regarding laterites, global distribution of laterites, and classification of laterites. The relationship between geology, slope forms, and slope processes in relation to laterites is interpreted by Townshend (1970). Paton and Williams (1972) successfully pointed out that the persistence of error in modern studies of laterite stemmed from early confusion over what laterite was and how it formed. They reviewed the early definitions of laterite, mentioning the tradition of the Wernerian School, Newbold School and Sedimentary Origin, Sedentary and detrital laterite, etc. Alongside the literatures of Alexander and Cady (1962), Persons (1970), Gidigasu (1976), Herbillon and Cedex (1988) and Aleva (1994) is mainly concerned with the new ideas and concepts in the geochemistry, morphology, and mineralogy of the lateritization processes with special emphasis on the tropical weathering profiles and formation laterites in different regions.

In the book, named '*Tropical Geomorphology*', Thomas (1974) had devoted one important chapter with special emphasis on laterites as a diagnostic weathering product of tropical region. He made detailed studies on laterite structure and morphology, classification of laterites, ideal profile of tropical laterites, occurrence of indurated laterites, development of lateritic formation on tropical terrain, origin of iron in laterites, genesis of secondary laterite formation, and landform development relating to laterite. Young (1976) referred laterite as a hard material, rich in secondary forms of iron. He concisely elaborated historical confusion about defining laterite, classification of laterite, composition and properties of laterite, climate and parent material, genesis of laterites, and landform evolution in relation to laterites.

The in-depth and field-based studies (in Uganda landscape of Africa) were done by McFarlane (1976) who had devoted eleven unique chapters (book entitled '*Laterites and Landscape*') regarding historical reviews of theories of laterite genesis, different views to define laterites, constituents of laterites, the environment of laterite with reference to geology, topography, climate, vegetation, profile of laterite, laterite structures, genesis of high-level and low-level laterites, lateritic landforms, and denudation chronology in Uganda. Another important book was introduced by Tardy (1997), entitled '*Petrology of Laterites and Tropical Soils*'. With 13 chapters the book focuses on the definition, distribution, arrangement, and formation of the ferruginous, aluminous or silico-aluminous hydrated or non-hydrated minerals within lateritic horizons, profiles, and landscapes of the present-day intertropical zone and ancient tropical palaeoenvironment.

Giving more stress on the difference between laterite and lateritic profile, some selected thoughts on laterite are provided by Eggleton (2001). They attempt to throw some light on the variable use of the word laterite, comment on laterite profile, briefly discussion about ferruginous accumulation with weathering profiles. Schaetzl and Anderson (2005) have presented some valuable study about pedogenic processes of humid and sub-humid tropics and they have more emphasized on formation of laterites, processes of laterization, latosolization, desilication, and rubification, and three phases of tropical pedogenesis. The literatures of Ollier and Galloway (1990), Tardy et al. (1991), Nahon and Tardy (1992), Tardy (1992), Bourman (1993, 1996), Pain and Ollier (1995) and Widdowson (2007, 2009) reveal the delineation and classification of laterites, variable processes of lateritization, precise chronology of onset of lateritization, identification of diagnostic signals of active lateritization, and geomorphic evolution of lateritic landscapes.

2.1.2   Indian Study

In India most of the in-depth studies regarding Indian laterites are done by eminent geologists. Roy Chowdhury et al. (1965) have revealed the origin of laterites in India, including classification and process of formation. Ray Chaudhuri (1980) has provided the concise account of Indian laterites and lateritic soils, including classification, description on laterites of West Bengal, and management of soils. Wadia (1999) provides few details on high-level and low-level laterites of India. In regional level, Hunday and Banerjee (1967) have studied the laterites of West Bengal and they have thought that these laterites are formed on fluviatile or estuarine Tertiary grits and gravels. The finest preliminary study on laterites and lateritic soils of West Bengal was done by Niyogi et al. (1970), Niyogi (1975), and Biswas (1987). Ghosh and Ghosh (2003) have written paper on laterites of Durgapur, Burdwan district in relation to its geological evolution and land degradation. Jha and Kapat (2003, 2009, 2011) have given notable contribution to the study of erosion prone laterites of Birbhum, West Bengal having spatial scale of C.D. block and mouza level. Similarly Das and Bandyopadhyay (1995) and Sen et al. (2004) have studied the laterites of Garhbeta, Paschim Medinipur. Chatterjee (2008) has done an excellent investigation on the description, characterization, and mapping of laterites in the Mayurakshi River Basin of Birbhum, West Bengal.

Many researchers and geoscientists of India and abroad have done finest works to understand the genesis, geochronology, geoarchaeology, landform evolution, and palaeogeography of lateritic terrain in the peninsular India. Among these, the literatures of Niyogi and Mallick (1973), Schmidt et al. (1983), Kumar (1986), Sychanthavong and Patel (1987), Sahasrabudhe and Rajaguru (1990), Devaraju and Khanadali (1993), Vaidyanadhan and Ghosh (1993), Achyuthan (1996, 2004) Widdowson and Cox (1996), Rajaguru et al. (2004), Mishra et al. (2007), Ollier and Sheth (2008), Chakraborti (2009), Bonnet et al. (2014, 2016), Ghosh and Gucchait (2015) Mathian et al. (2019) and Liu et al. (2019) are worth to be mentioned (Table 2.1 and Fig. 2.1).

Most of these works affirm that laterite profiles are used as the weathering archive and the storehouse of palaeoclimatic information. The extensive palaeomagnetic

Table 2.1 Important research reports, books, book chapters, and articles on Laterites (at international and national level) considered in this study

Authors	Theme/Title
Paton and Williams (1972)	The concept of laterites
Niyogi et al. (1970)	A preliminary study of laterites of West Bengal, India
Biswas (1987)	Laterites and lateritoids of Bengal
McFarlane (1976)	Laterites and landscape
Scmidt et al. (1983)	Magnetic ages of some Indian laterites
Kumar (1986)	Palaeolatitudes and the age of Indian laterites
Roy Chowdhury (1986)	Concepts and origin of Indian laterite in historical perspective
Sychanthavong and Patel (1987)	Laterites and lignites of north-western India and their relevance to the drift tectonics of the Indian Plate
Tardy et al. (1991)	Mineralogical composition and geographical distribution of African and Brazilian periatlantic laterites—the influence of continental drifty and tropical paleoclimates during the past 150 million years and implications for India and Australia
Tardy (1992)	Diversity and terminology of laterite profile
Bourman (1993)	Perennial problems in the study of laterite: a review
Pain and Ollier (1995)	Inversion of relief—a component of landscape evolution
Widdowson and Cox (1996)	Uplift and erosional history of the Deccan Traps, India: evidence from laterites and drainage patterns of the Western Ghats and Kankan Coast
Tardy (1997)	Petrology of laterites and tropical soils
Ollier and Sheth (2008)	The High Deccan duricrust of India and their significance for the 'laterite' issue
Bonnet et al. (2016)	Cenozoic lateritic weathering and erosion history of Peninsular India from $^{40}Ar/^{39}Ar$ dating of supergene K–Mn oxides
Mathian et al. (2019)	Unravelling weathering episodes in Tertiary regoliths by kaolinite dating (Western Ghats, India)
Liu et al. (2019)	Aeolian accumulation: an alternative origin of laterite of the Deccan Plateau, India

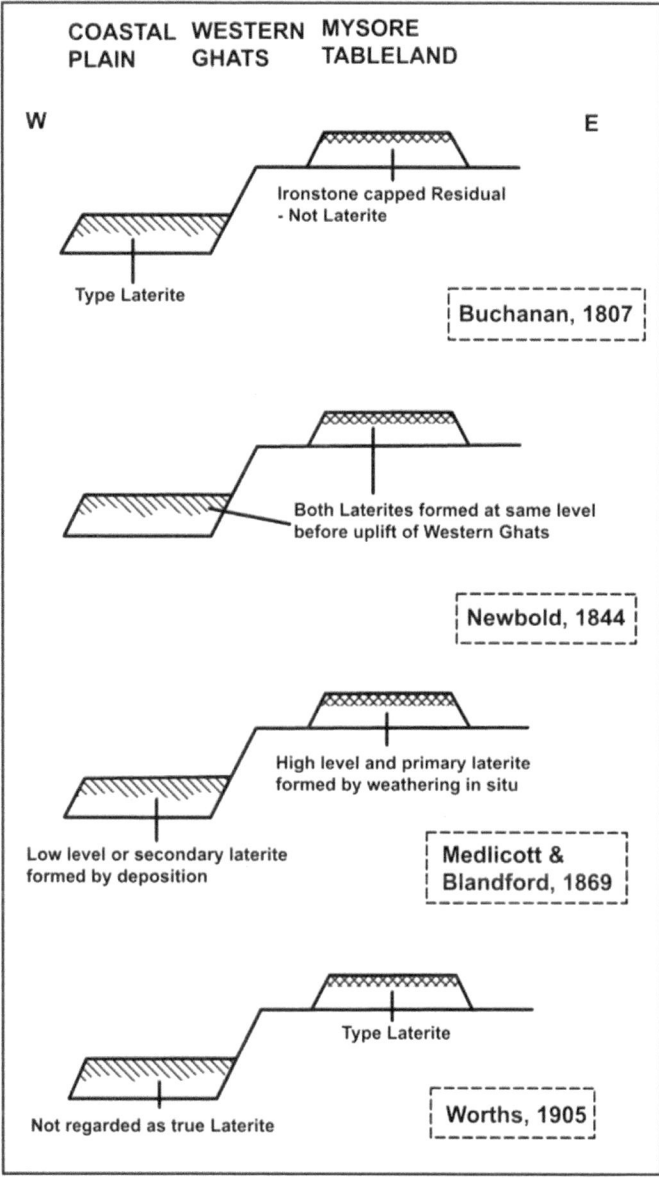

Fig. 2.1 Changing concepts of laterites in India (Paton and Williams 1972)

investigation of laterites is described that the topographically high-level laterites have undergone a complex magnetic history during the Late Cretaceous–Early Tertiary and the low-level laterites are Mid–Late Tertiary in age (Schmidt et al. 1983; Kumar 1986). Dating of laterite samples reveals a period of intense tropical weathering in Late Eocene–Oligocene with three successive phases of in situ lateritic weathering optimum in India—(1) Early Eocene, (2) Late Eocene–Late Oligocene, and (3) Early Miocene (Bonnet et al. 2016). Recently dating using electron paramagnetic resonance spectroscopy has confirmed that four major lateritic weathering episodes (~1–39 Ma) occurred lining with the strengthening of south-west monsoon (Mathian et al. 2019). Ollier and Sheth (2008) have applied a theory of inversion of relief to the laterites of peninsular India, but recently Liu et al. (2019) confirm the theory of aeolian accumulation to understand the origin of laterites.

2.2 Research Gaps and Objectives

The laterite described by Buchanan (1807) is only one member of laterite families whose members have different properties but similar genesis (Fig. 2.1). Laterites have raised controversy ever since the term was coined and disagreements over the definition of laterite have continued sporadically for 150 years (Ghosh and Guchhait 2019). There has been renewal interest in the geochemistry, genesis, and formation of laterites during the International Geological Correlation Programme (IGCP) project on 'Lateritization Processes' held in Trivandrum, India (1979). In spite of numerous publications and researches on laterites, much confusion, contradiction, and controversies still proliferate in the available literatures on the genesis, distribution, classification, geological age, subsurface profiling of laterites, and present-day lateritization process (Ghosh and Guchhait 2019).

Studies of long timescales of landscape evolution are often considered less important in the geomorphology and Quaternary studies than those dealing now with processes and chronologically secure high-resolution sequences of Late Quaternary age (Rajaguru et al. 2004). But this can be false dichotomy, as any landscape is a product of forms, materials of different origins and ages and the present-day landscape has roots in relict and ancient environment with overprint of the Quaternary processes (Sahasrabudhe and Rajaguru 1990; Rajaguru et al. 2004; Ghosh and Guchhait 2019). The laterites of India are such geological unit which usually reflects relict-weathered products of pre-Anthropocene palaeoenvironment, morpho-stratigraphic markers of past and unique paleogeomoprhic evolution since Palaeogene. Laterites reflect a past geo-climatic environment which is not present in India and as well as in West Bengal, because many researchers agreed that existing laterites are clearly relics of geological antiquity (Ghosh and Guchhait 2019). The time span and ideal climate needed to create a fully developed laterite profiles is still unknown to many researchers.

There are good and significant researches on the laterites of peninsular India as described earlier, but there are important research gaps to investigate the laterites of Bengal Basin. To understand genesis and development of ferruginous laterites (mainly in *Rarh* Plain) some imperative considerations, quires, and research needs are born in mind (Ghosh and Guchhait 2019).

(1) There is a question regarding source of ferralitic material which contributed to make up reddish brown duricrust of *Rarh*.
(2) Whether laterite and its variant are highly related to tropical chemical weathering of RBT and other rocks or western shelf deposits of Bengal Basin.
(3) Whether primary (in situ) or secondary (ex situ) origin of laterites (i.e. modes of formation) is observed in the region.
(4) Whether there is any possibility or potentiality of these laterites as an indicator of palaeoclimate and palaeogeomophic event (simultaneously tectono-climatic evolution).
(5) A key issue is raised about the exact age and geochronology of laterites in this region.
(6) There is a research need to understand the relation between drifting of Indian plate and establishment of lateritization climate.

Above all one basic question is borne in mind—what are reflected from the laterites to known about past environments and events? Understanding above research needs and question, this work has set forth three prime objectives within the domain of regolith geology and palaeogeography.

- Analysing the lithofacies of laterites to evaluate in situ and ex situ origin,
- Determining the possible time span of lateritization event, and
- Exploring the palaeoclimatic and palaeogeomorphic significance of laterites.

2.3 Research Methodology

2.3.1 Sample Sites

The combined zones of primary and secondary laterites with Neogene gravel deposits are the main spatial unit of study, bounded by the latitude of 21° 30′ to 24° 40′ N and longitude of 86° 45′ to 87° 50′ E (Fig. 2.2). Geologically this part of West Bengal was formerly developed as the stable shelf province of Bengal Basin which experienced severally marine regression and transgression since Miocene, related to climate change and neo-tectonic activity (Alam et al. 2003). Fringing the basalts of Early Cretaceous Rajmahal Basalt Traps, sandstones of Gondwana Formations (Permian–Silurian) and granite–gneiss of Archaean Formation at west and Late Quaternary Older Alluvium (Panskura and Sijua Formations) at east, the lateritic zone of *Rarh Bengal* is found at the north-western border of Bengal Basin. Detailed field investigations of weathering archives are conducted in the sample sites (covering study area of *Rarh*) of Bataspur (23° 55′ 47″ N, 87° 29′ 17″ E), Sultanpur (24° 22′ 33″ N, 87° 47′ 31″ E), Rautara (23° 53′ 12″ N, 87° 28′ 15″ E), Sukna (23° 58′ 55″ N, 87°

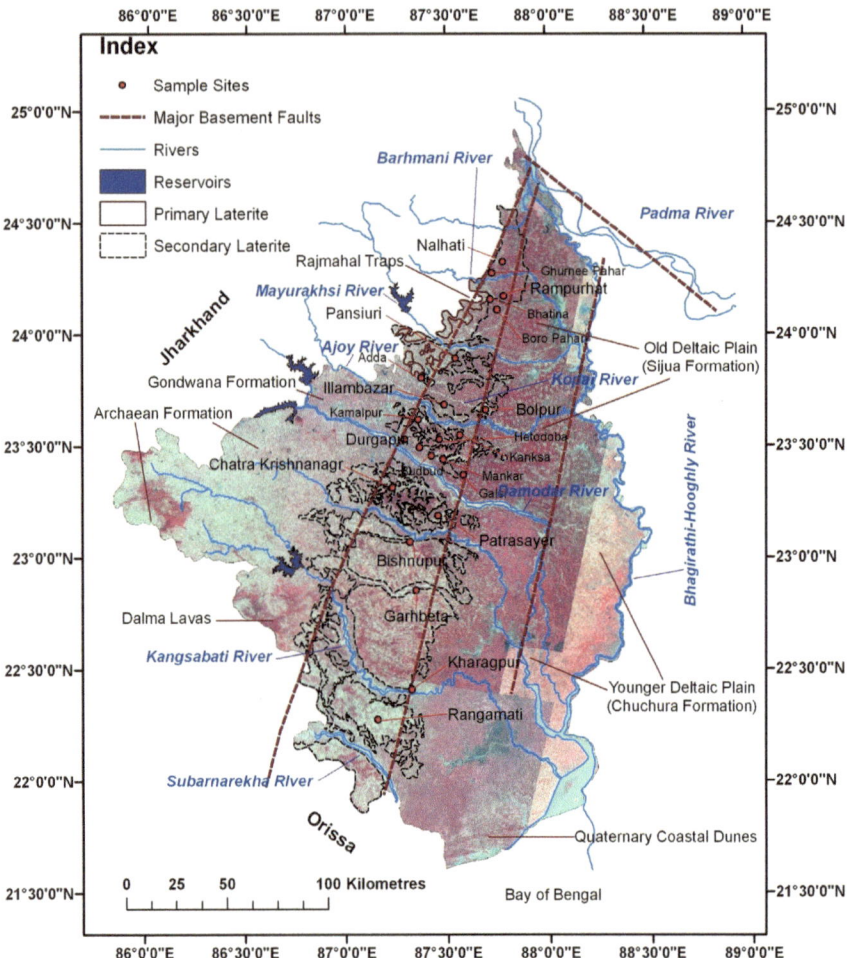

Fig. 2.2 Laterite exposures of Bengal Basin in relation to basin margin faults (Ghosh and Gucchait 2015)

30′ 52″ E), Husennagar (23° 48′ 04″ N, 87° 08′ 34″ E), Nalhati (24° 17′ 47″ N, 82° 49′ 28″ E), Pinargaria (24° 12′ 13″ N, 87° 40′ 13″ E), Ghurnee Pahar (24° 15′ 43″ N, 87° 39′ 11″ E), Bastabpur (23° 48′ 50″ N, 87° 13′ 29″ E), Hetampur (23° 47′ 03″ N, 87° 23′ 37″ E), Deucha (23° 24′ 06″ N, 87° 14′ 55″ E), Dubrajpur (23° 47′ 12″ N and 87° 25′ 19″ E), Barjora (23° 25′ 41″ N, 87° 16′ 32″ E), Sriniketan (23° 41′ 31″ N and 87° 40′ 31″ E), Hetodoba (23° 26′ 45″ N and 87° 32′ 07″ E), Bishnupur (23° 05′ 28″ N and 87° 16′ 15″ E), Garhbeta (23° 05′ 28″ N and 87° 16′ 15″ E), Radha Damodarpur (23° 04′ 15″ N, 87° 22′ 23″ E), Gangani (22° 51′ 21″ N, 87° 20′ 45″ E), Nanda Bhanga (23° 10′ 31″ N, 87° 15′ 51″ E), Durgadaspur (22° 32′ 00″ N, 87° 19′ 10″ E), Jamsol (22° 25′ 54″ N, 87° 14′ 22″ E), Kharagpur (22° 17′ 37″ N, 87°

15′ 10″ E), Illambazar (23° 36′ 53″ N, 87° 32′ 00″ E), Surul (23° 40′ 18″ N, 87° 38′ 52″ E), Kharia (23° 59′ 11″ N, 87° 35′ 31″ E), and Rangamati (22° 24′ 42″ N, 87° 17′ 55″ E), covering the span of study area.

2.3.2 Secondary Data Collection

To collect spatial information, the topographical sheets (1 : 50, 000 scale) of Survey of India (SOI, 72 P/12, 73 M/3, M/4, M/6, M/10, M/11, and 73 N/1) and GSI (Geological Survey of India) district resource maps of Birbhum, Bardhaman, Bankura, and West Medinipur districts are severally used in research. The outline map of *Rarh Bengal* is prepared in ArcGis 9.1 with the help of a base-map prepared by Bagchi and Mukherjee (1983). With the help of shapefile format, we have prepared thematic maps using option of subset in Erdas 9.1. To identify the lateritic zone, we have applied NDVI (Normalized Difference Vegetation Index) and Iron Oxide Index in Erdas 9.1 imagine using post-monsoon GLCF (Global Land Cover Facility, http://glcf. umd.edu/data/landsat/) Landsat ETM+ images of 2000–2001 (path/row—138/43, 138,44, 138/45, 139/43, 139/44, 139/45, and 140/44). The ASTER (Advanced Space Borne Thermal Emission and Radiometer) elevation data with 30 m resolution of 2011 is collected from the website of Earth Explorer (http://earthexplorer.usgs.gov/). Alongside we have used the unpublished geological expedition reports of GSI, Eastern Region which are regularly provided in the official website (http://www.portal. gsi.gov.in). The total workflow of this work is depicted in flow chart (Fig. 2.3).

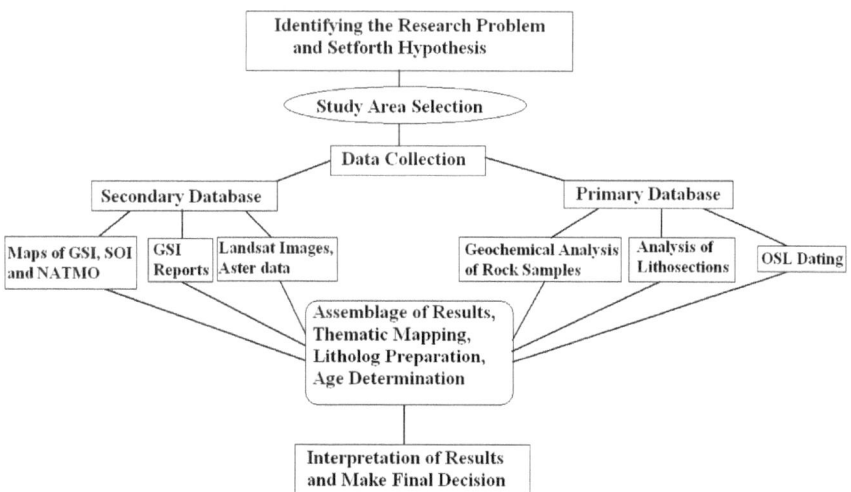

Fig. 2.3 Flow chart of adopted methodology used in this study (Ghosh and Guchhait 2019)

2.3.3   Geochemical Analysis

The lithosections of laterite profiles are investigated and analysed on the basis of texture, colour, cementation, degree of mottling and bleaching, weathering front, iron–aluminium oxides assemblages and other petrographic and geochemical properties. The chemical properties of laterite samples belonging to dismantled duricrust, ferricrete, clay horizons, and saprolite zones are analysed along the profile with depth to know the variable percentages of SiO_2, Al_2O_3, Fe_2O_3, MnO, MgO, CaO, Na_2O, K_2O, TiO_2, and P_2O_5. Then the molar ratio SiO_2/R_2O_3 (where $R_2O_3 = Al_2O_3 + Fe_2O_3 + TiO_2$), suggested as an index of tropical weathering (Birkeland 1984; Singh et al. 1998), is used in the sample sites for increasing degree of leaching and Fe–Al segregations at the top. To know the degree of lateritization, we have used Al_2O_3–Fe_2O_3–SiO_2 triangular diagram (Schellman 1986; Pain and Ollier 1995; Meshram and Randive 2011) in this study. Chemical weathering strongly affects the mineralogy and major elements geochemistry of the bedrocks. The degree of weathering can be evaluated by quantitative measures using whole rock chemical analysis such as Chemical Index of Alteration (CIA) defined by Nesbitt and Young (1982) and Meshram and Randive (2011).

$$CIA = 100\, Al_2O_3/(Al_2O_3 + CaO + Na_2O + K_2O)$$

2.3.4   OSL Dating Method of Lithofacies

The choice of a particular proxy depends on the period for which palaeoclimate information is being retrieved, the site and availability of suitable material for extracting the proxy data. In general, any dating method can be understood by a simple example of a beaker being filled at a certain input drop rate or being emptied at a known rate (Singh et al. 1998; Singhvi and Kale 2009). Out of a number of dating tools, radiocarbon dating (^{14}C) is historically the most commonly used technique, but ^{14}C dating is limited to the environments where organic matter is preserved. As an alternative, optically stimulated luminescence (OSL) dating is applied here to find the accurate age of facies. OSL dating method uses minerals (quartz and feldspar) that are common in most environments making the technique applicable in the tropical landscapes of laterites. Fluvial sediments and fan deposits are the natural archives which have been used here in palaeoclimatic reconstruction (Ghosh and Guchhait 2014). The isotopic composition and mineralogical changes are the recommended climate proxies of various laterite profiles (i.e. primary and secondary laterites) which enable quantification of the change in climate in the past (Ghosh and Guchhait 2014).

Luminescence dating determines the age since burial of sediment by measuring the total amount of stored signal resulting from exposure of the sediment to a known annual dose of background radiation (Aitken 1998; Duller 2004; Preusser et al. 2008). OSL was introduced for dating by Huntley et al. (1985) and in geological dating, it is most important because the trapped electron component is most likely to be emptied during transport prior to deposition and burial. The precise method of luminescence dating was reviewed, elaborated, and used by Aitken (1998), Prescott and Robertson (1997), Wintle (1997, 2008), Stokes (1999), Murray and Olley (2002), Duller (2004), Preusser et al. (2008), and Rittenour (2008). This dating method is successfully applied in the practical fields of geology, geomorphology, archaeology, and other palaeogeographic studies in India (Singhvi et al. 1982; Sankaran et al. 1985; Singh et al. 1998; Kale et al. 2000; Briant et al. 2006; Sridhar 2007; Thomas et al. 2007; Singhvi and Porat 2008; Singhvi and Kale 2009; Rajaguru et al. 2011; Metha et al. 2012; Ghosh and Guchhait 2019).

There is almost no research work to detect the age of lateritization in West Bengal using advanced dating methods. Here to unearth the geochronology of *Rarh* laterite and its palaeogeographic importance, the Geological Survey of India (Chakraborti 2011) have effectively used OSL dating method using well-known Risø TL/OSL-DA-15 reader using internal Sr/Y-90 beta source and a combination of Schott UG 11 and BG-39 filters. The detailed measurements of luminescence characteristics and equivalent dose distributions in nine samples of Sriniketan, Bishnupur, and Garhbeta (representative samples of *Rarh* laterites) are key elements of this research work. The key element for the success of OSL dating method is the validity of the assumption of closed system, i.e. the system had no unknown inputs or leakages during the time period under consideration and that the drop or leak rate remained constant throughout and, if they did change then the magnitude and the style of change through time was exactly known (Ghosh and Guchhait 2019).

The energy stored increases with the amount of radiation to which a crystal is exposed and this provides a 'clock' that is the basis of all luminescence dating methods (Duller 2004; Thomas et al. 2007). Typically, irradiation at ~500 nm/2.5 eV leads to emission at wavelength shorter than 400 nm/3.1 eV which can be isolated from the stimulating wavelength by optical filters (Prescott and Robertson 1997). This is commonly known as OSL which introduced for dating by Huntley et al. (1985). Luminescence dating determines the age since burial of sediment by measuring the total amount of stored signal resulting from exposure of the sediment to a known annual dose of background radiation. The concept of luminescence dating relies on defects in the crystal lattice of dosimeter minerals, most commonly quartz and feldspar, to trap energy produced during the interaction between electrons within the crystal and background radiation from the radioactive decay of uranium (U), thorium (Th) and potassium (K), and cosmic rays (Singh et al. 1998; Singhvi and Porat 2008; Singhvi and Kale 2009).

The principle is expressed in the 'age' equation (Aitken 1998), where equivalent dose is the radiation dose delivered to the mineral grains in the laboratory to stimulate luminescence and dose rate is the rate at which ionizing energy is delivered from the background radiation. The acquired thermoluminescence at any time is related to the time elapsed since the burial. The basic equation thermoluminescence age is the ratio between total acquired thermoluminescence since burial and rate of thermoluminescence acquisition (Singh et al. 1998; Preusser et al. 2008; Singhvi and Kale 2009). The SI unit for dose is Gray (Gy) which is a measure of how much energy is absorbed by a sample in joules per kilogram ($J\ kg^{-1}$). The dose rate computation assumed a radioactive decay series. The measurement errors using the conventional error calculation method in these cases are computed to be 10–12% and a working estimate of total error should be taken as $\pm 15\%$. Routine OSL dating of quartz can give ages from 10 ± 3 ka to 150 ka (Stokes 1999; Duller 2004). Fluvial sediments and weathered lithologies, commonly dated as bleaching regimes in this environment, are relatively well understood. It has been demonstrated that when gravels are dominated by inert lithologies, sand within gravel matrix provides here reliable age estimates (Ghosh and Guchhait 2019).

References

Achyuthan H (1996) Geomorphic evolution and genesis of laterites around the east coast of Madras, Tamil Nadu, India. Geomorphology 16:71–78

Achyuthan H (2004) Paleopedology of ferricrete horizons around Chennai, Tamil Nadu, India. Revista Mexicana de Ciencias Geologicas 21(1):133–143

Aitken MJ (1998) An introduction to optical dating. Oxford University Press, New York

Alam M, Alam MM, Curray JR, Chowdhary MLR, Gandhi MR (2003) An overview of the sediment geology of the Bengal Basin in relation to the regional tectonic framework and basin-fill history. Sediment Geol 155(3–4):179–208

Aleva GJJ (1994) Laterites—concepts, geology, morphology and chemistry. ISRIC, Wageningen

Alexander LT, Cady JG (1962) Genesis and hardening of laterite in soils. United States Department of Agriculture. Tech Bull 1281:1–90

Bagchi K, Mukherjee KN (1983) Diagnostic survey of Rarh Bengal (Part II). University of Calcutta, Calcutta

Birkeland PW (1984) Soils and geomorphology. Oxford University Press, New York

Biswas A (1987) Laterities and lateritoids of Bengal. In: Datye VS, Diddee J, Jog SR, Patial C (eds) Exploration in the tropics. K.R. Dikshit Felicatiobn Committee, Pune, pp 157–167

Bonnet NJ, Beauvais A, Arnaud N, Chardon D, Jayananda M (2014) First $^{40}Ar/^{39}Ar$ dating of intense Late Palaeogene lateritic weathering in peninsular India. Earth Planet Sci Lett 386:126–137

Bonnet N, Beauvais A, Arnaud N, Charden D, Jayananda M (2016) Cenozoic lateritic weathering and erosion history of peninsular India from $^{40}Ar/^{39}Ar$ dating of supergene K-Mn oxides. Chem Geol 446:33–53

Bourman RP (1993) Perennial problems in the study of laterite: a review. Aust J Earth Sci 40(4):387–401

Bourman RP (1996) Towards distinguishing transported and in-situ ferricretes: data from southern Australia. J Aust Geol Geophys 16(3):231–241

Briant RM, Bates MR, Schwenninger J, Wenban-Smith F (2006) An optically stimulated luminescence dated Middle to Late Pleistocene fluvial sequence from the western Soloent Basin, southern England. J Quater Sci 24(8):916–927

Buchanan F (1807) A journey from Madras through the countries of Mysore, Kanara and Malabar (3 volumes). East India Company, London

Chakraborti S (2009) Quaternary laterites of West Bengal. NEWS Geol Surv India 30(1/2):11–12

Chakraborti S (2011) Final report on Quaternary laterites in the western districts of West Bengal—their geomorphology, stratigraphy, genesis and implications for climate change. Geological Survey of India Eastern Region, Kolkata, pp 1–88

Chatterjee N (2008) Laterite terrains of the Chotanagpur Plateau fringe region (case study of the Mayurakshi Basin, eastern India). Indian J Landsc Syst Ecol Stud 31(1):115–130

Das K, Bandyopadhyay S (1995) Badland development over laterite duricrusts. In: Jog SR (ed) Indian Geomorphology, vol I. Erosional landforms and processes. Rawat Publication, New Delhi, pp 31–41

Devaraju TC, Khanadali SD (1993) Laterite bauxite profiles of south western and southern India—characteristics and tectonic significance. Curr Sci 64(11–12):919–921

Duller GAT (2004) Luminescence dating of Quaternary sediments: recent advances. J Quat Sci 19(2):183–192

Eggleton RA (2001) The regolith glossary. CRC Australia CRC LEME, Canberra

Ghosh S, Ghosh S (2003) Land degradation due to indiscriminate Murrum extraction near Durgapur Town, West Bengal. In: Jha VC (ed) Land degradation and desertification. Rawat Publications, Jaipur, pp 255–267

Ghosh S, Guchhait SK (2014) Palaeoenvironmental significance of fluvial facies and archives of Late Quaternary deposits in the floodplain of Damodar River, India. Arabian J Geosci 7(10):4145–4161

Ghosh S, Guchhait S (2015) Chraterization and evolution of primary and secondary laterites in northwestern Bengal Basin, West Bengal, India. J Palaeogeogr 4(2):203–230

Ghosh S, Guchhait SK (2019) Modes of formation, Palaeogene to Early Quaternary Palaeogenesis and geochronology of laterites in Rajmahal Basalt Traps and Rarh Bengal of Lower Ganga Basin. In: Das BC, Ghosh S, Islam A (eds) Quaternary geomorphology in India. Springer, Singapore, pp 25–60

Gidigasu MD (1976) Laterite soil engineering: pedogenesis and engineering principles. Elsevier, Amsterdam

Herbillon AJ, Cedex V (1988) Laterites and lateritization processes. In: Stucki JW et al (eds) Iron in soils and clay minerals. D. Reidel Publishing Company, London, pp 779–796

Hunday A, Banerjee S (1967) Geology and mineral resources of West Bengal. Memoirs of the geological survey of India, Delhi

Huntley DJ, Godfrey-Smith DI, Thevalt MLW (1985) Optical dating of sediments. Nature 313:105–107

Jha VC, Kapat S (2003) Gully erosion and its implications on land use, a case study. In: Jha VC (ed) Land degradation and desertification. Rawat Publication, Jaipur, pp 156–178

Jha VC, Kapat S (2009) Rill and gully erosion risk of laterite terrain in south western Birbhum district, West Bengal, India. Sociedade & Natureza Uberlandia 21(2):141–158

Jha VC, Kapat S (2011) Degraded lateritic soilscape and land uses in Birbhum district, West Bengal, India. Sociedade & Natureza Uberlandia 23(3):545–558

Kale VS, Singhvi AK, Mishra PK, Banerjee D (2000) Sedimentary records and luminescence chronology of Late Holocene palaeo-floods in the Luni River, Thar Desert, northwest India. CATENA 40:337–3587

Kumar A (1986) Palaeolatitudes and the age of Indian laterites. Palaeogeogr Palaeoclimatol Palaeoecol 53:231–237

Liu X, Mao X, Yuan Y, Ma M (2019) Aeolian accumulation: an alternative origin of laterite on the Deccan Plateau, India. Palaeogeogr Palaeoclimatol Palaeoecol 518:34–44

Maignien R (1966) Review of research on laterites. UNESCO, Paris

Mathian M, Aufort J, Braun J, Riotte J, Selo M, Balan E, Fritsch E, Bhattacharya S, Allard T (2019) Unraveling weathering episodes in Tertiary regoliths by kaolinite dating (Western Ghats, India). Gondwana Res 69:89–105

McFarlane MJ (1976) Laterite and landscape. Academic Press, London

Meshram RR, Randive KR (2011) Geochemical study of laterites of the Jamnagar district, Gujarat, India: implications on parent rock, mineralogy and tectonics. J Asian Earth Sci 42:1271–1287

Metha M, Majeed Z, Dobhal DP, Srivastava P (2012) Geomorphological evidences of post—LGM glacial advancements in the Himalaya: a study from Chorabari Glacier, Garhwal Himalaya, India. J Earth Syst Sci 121(1):149–163

Mishra S, Deo S, Rajaguru SN (2007) Some observations on the laterites developed on Deccan Trap: implications for the Post-Deccan Trap denudational history. J Geol Soc India 70:469–475

Murray AS, Olley JM (2002) Precision and accuracy in the optically stimulated luminescence dating of sedimentary quartz: a status review. Geochronometria 21:1–16

Nahon D, Tardy Y (1992) The ferruginous laterites. In: Butt CRM, Zeeges H (eds) Regolith exploration geochemistry in tropical and subtropical terrain. Elsiver, Amsterdam, pp 41–55

Nesbitt HW, Young GM (1982) Early proterozoic climates and plate motion inferred from major element chemistry of lutites. Nature 299:715–717

Niyogi D (1975) Quaternary geology of the coastal plain in West Bengal and Orissa. Indian J Earth Sci 2:51–61

Niyogi D, Mallick S (1973) Quaternary laterites of West Bengal: its morphology, stratigraphy and genesis. Quat J Geol Min Metall Soc India 45(4):157–174

Niyogi D, Mallick S, Sarkar SK (1970) A preliminary study of laterites of West Bengal, India. In: Chatterjee SP, Das Gupta SP (eds) Selected papers physical geography (vol 1). 21st international geographical congress, Calcutta, National Committee for Geography, pp 443–449

Ollier CD, Galloway RW (1990) The laterite profile, ferricrete and unconformity. CATENA 17:97–109

Ollier CD, Sheth HC (2008) The high Deccan duricrusts of Indian and their significance for the 'laterite' issue. J Earth Syst Sci 117(5):537–551

Pain CF, Ollier CD (1995) Inversion of relief—a component of landscape evolution. Geomorphology 12:151–165

Paton TR, Williams MAJ (1972) The concept of laterite. Ann Assoc Am Geogr 62(1):42–56

Persons BS (1970) Laterite—genesis, location, use. Plenum Press, New York

Prescott JR, Robertson GB (1997) Sediment dating by luminescence: a review. Radiat Meas 27(5–6):893–922

Preusser F, Degering D, Fuchs M, Hilgers A, Kadereit A, Klasen N, Krbetschek M, Richter D, Spencer JQG (2008) Luminescence dating: basics, methods and applications. Quat Sci J 57(1–2):95–149

Rajaguru SN, Deo SG, Mishra S, Ghate S, Naik S, Shirvalkar P (2004) Geoarchaeological significance of the detrital laterites discovery in the Karha Basin, Pune District, Maharastra. Man Environ XXIX(1):1–6

Rajaguru SN, Deotare BC, Gangopadhyay K, Sain MK, Panja S (2011) Potential geoarchaeological sites for luminescence dating in the Ganga Bhagirathi-Hugli delta, West Bengal, India. Geochronometria 38(3):282–297

Ray Chaudhuri SP (1980) The occurrence, distribution, classification and management of laterite and laterite soils. Cahiers O.R.S.T.O.M. Serie Pedologie 18(3–4):249–252

Rittenour TM (2008) Luminescence dating of fluvial deposits: applications to geomorphic, palaeoseismic and archaeological research. Boreas 37:613–635

Roy Chowdhury MK, Venkatesh V, Anandalwar MA, Paul DK (1965) Recent concepts on the origin of Indian laterite. Proc Natl Acad Sci India Sect A Phys Sci A 31(6):547–558

Roy Chowdhury MK (1986) Concepts of the origin of Indian laterite in historical perspective. Proceedings of the national institute of science India, 52A(6), pp 1307–1323

Sahasrabudhe YS, Rajaguru SN (1990) The laterites of Maharashtra State. Bull Deccan Coll Res Inst 49:357–370

Sankaran AV, Nambi KSV, Sunta CM (1985) Thermoluminescence of laterites: applicability in dating. Nucl Tracks 17(5):177–183

Schaetzl RJ, Anderson S (2005) Soils: genesis and geomorphology Cambridge University Press, Cambridge

Schellmann W (1986) A new definition of laterite. In: Banerjee PK (ed) Lateritisation processes. Geological Survey of India Memoir, vol 120, pp 11–17

Schmidt PW, Prasad V, Raman PK (1983) Magnetic ages of some Indian laterites. Palaeogeogr Palaeoclimatol Palaeoecol 44:185–202

Sen J, Sen S, Bandyopadhyay S (2004) Geomorphological investigations of badlands: a case study at Garhbeta, west Medinipur district, West Bengal, India. In: Singh S, Sharma HS, De SK (eds.) Geomorphology and environment. ACB Publication, Kolkata, pp 204–234

Singh LP, Parkash B, Singhvi AK (1998) Evolution of the Lower Gangetic Plain landforms and soils in West Bengal, India. CATENA 33:75–104

Singhvi AK, Porat N (2008) Impact of luminescence dating on geomorphological and palaeoclimate research in drylands. Boreas 37(4):536–558

Singhvi AK, Kale VS (2009) Paleoclimate studies in India: last ice age to the present. IGBP WCRP SCOPE Rep Ser 4:1–28

Singhvi AK, Sharma YP, Agrawal DP (1982) Thermo-luminescence dating of sand dunes in Rajasthan. Nature 295:313

Stokes S (1999) Luminescence dating applications in geomorphological research. Geomorphology 29:153–171

Sychanthavong SPH, Patel PK (1987) Laterites and lignites of northwestern India and their relevance to the drift tectonics of the Indian Plate. Curr Sci 56(10):469–473

Tardy Y (1992) Diversity and terminology of laterite profile. In: Martini IP, Chesworth W (eds) Weathering, soils and paleosols. Elsevier, Amsterdam, pp 379–405

Tardy Y (1997) Petrology of laterites and tropical soils. AA Bakema, Rotterdam

Tardy Y, Kobilsex B, Paquet H (1991) Mineralogical composition of geographical distribution of African and Brazilian peri-Atlantic laterites: the influence of continental drift and tropical paleoclimates during the past 150 million years and implications for India and Australia. J Afr Earth Sci 12(1–2):283–295

Thomas MF (1974) Tropical geomorphology: a study of weathering and landform development in warm climates. MacMillan Press Ltd., New York

Thomas PJ, Juyal N, Kale VS, Singhvi AK (2007) Luminescence chronology of Late Holocene extreme hydrological events in the upper Pennar River Basin, south India. J Quater Sci 22(8):747–753

Townshend JRG (1970) Geology, slope form and slope process and their relation to the occurrence of laterite. Geogr J 136(3):392–399

Vaidyanadhan R, Ghosh RN (1993) Quaternary of the east coast of India. Curr Sci 31(6):231–232

Wadia DN (1999) Geology of India. Tata McGraw Hill, New Delhi

Widdowson M (2007) Laterite and ferricrete. In: Nash DJ, McLaren SJ (eds) Geochemical sediments and landscapes. Blackwell Publsihing, Maiden, pp 46–94

Widdowson M (2009) Laterite. In: Tegen I, Gornitz V (eds) Encyclopedia of paleoclimatology and ancient environments. Springer, Netherlands, pp 514–517

Widdowson M, Cox KG (1996) Uplift and erosional history of the Deccan Traps, India: evidence from laterites and drainage patterns of the Western Ghats and Kankan Coast. Earth Planet Sci Lett 137:57–69

Wintle AG (1997) Luminescence dating: laboratory procedures and protocols. Radiat Meas 27(5–6):769–817

Wintle AG (2008) Luminescence dating: where it has been and where it is going. Boreas 37:471–482

Young A (1976) Tropical soils and soil survey. Cambridge University Press, Cambridge

Chapter 3
Geographical Settings of Study Area

Abstract Laterites of Bengal Basin are linked with the historical evolution of the Basin itself and ferruginization of lateritic sediments is linked with many geomorphic phenomena since Palaeocene. The lateritic region of West Bengal is popularized as '*Rarh Bengal*' which is the uplifted interfluves and parts of palaeo-deltas (much older than Ganga–Brahmaputra Delta), formed by east flowing peninsular rivers. This section tries to unearth the tectonic–geomorphic evolution of Bengal Basin and its geological structure which are essential and fundamental part of the study, because the platform of laterite genesis is connected with the Basin neotectonics. Alongside to understand the overall geographic features, the climate, soils, and natural vegetation are discussed here.

Keywords Geological structure · Neotectonics · Chota Nagpur Plateau · Bengal Basin · Ganga–Brahmaputra–Meghna Delta

3.1 Occurrence of Laterites in Study Area

The many countries of world (mainly the continents of Africa, India, Australia, etc.), which passed through intense tropical wet-dry climate during post-Cretaceous continental drift, have extensive glimpse of red-brown ferruginous crust in surface or near-surface of earth. The tropical to subtropical wet-dry types of climate, mainly 'rubefaction zone' of Pedro (1968), are allied with the ferruginous crusts which are widely recognized in India as laterites or ferricretes or plinthites (Pendleton 1936; Pascoe 1964; McFarlane 1976; Tardy 1992; Bourman 1993). In West Bengal, the lateritic uplands or upland red soil groups (Singh et al. 1998) occur along a NE–SW trending belt parallel to the western margin of Bengal Basin. This unique geomorphic region (i.e. shelf zone of Bengal Basin) is designated as *Rarh Bengal* by Bagchi and Mukherjee (1983). The duricrusts, ferruginous gravels, and kaolinite deposits (from Rajmahal Basalt Traps to Subarnarekha Basin) are appeared as interfluves and borders this province to make the transitional diagnostic tropical weathering surface and distinct sedimentary lithofacies in between the Archaean–Gondwana litho-unit

© The Author(s), under exclusive license to Springer Nature Switzerland AG 2020   33
S. Ghosh and S. K. Guchhait, *Laterites of the Bengal Basin*,
SpringerBriefs in Geography, https://doi.org/10.1007/978-3-030-22937-5_3

at west and the Quaternary alluvial litho-unit of Bengal Basin at east (Niyogi et al. 1970; Niyogi 1975; Biswas 1987; Das Gupta and Mukherjee 2006).

3.1.1 Study Area

The zones of laterites with Neogene gravel deposits are the main spatial unit of study, bounded by the latitude of 21° 30′ to 24° 40′ N and longitude of 86° 45′ to 87° 50′ E (Fig. 3.1). Geomorphologically this part of West Bengal is recognized as western fringe of the Ganga–Brahmaputra–Meghna (GBM) Delta and geologically this

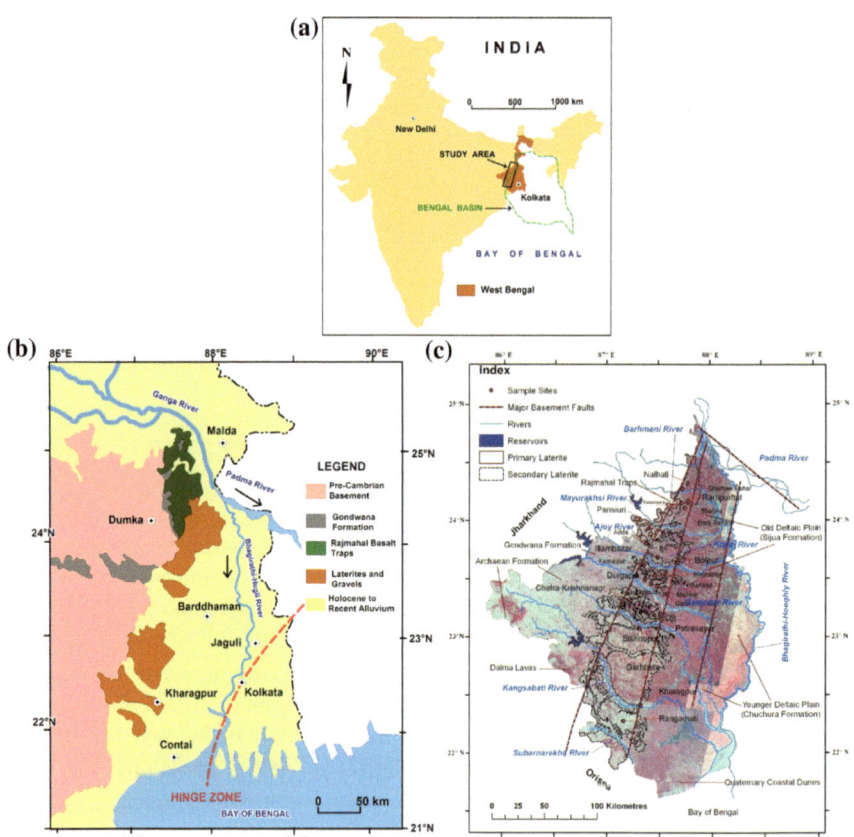

Fig. 3.1 a Location of study area and Bengal Basin in eastern part of India, **b** general surface geology of north-west part of Bengal Basin (modified from Das Gupta and Mukherjee 2006), and **c** identified units of primary and secondary laterites (i.e. study area) with basement fault system in West Bengal, dissected by west–east flowing parallel rivers and merged with Older Deltaic Plain (Sijua Formation) of Bengal Basin at east (Ghosh and Guchhait 2019)

part was formerly developed as the western stable shelf province of the Bengal Basin which experienced severally marine regression and transgression since Miocene, related to climate change and neo-tectonic activity (Alam et al. 2003; Ghosh and Guchhait 2015, 2019). The distribution of laterites and ferruginous soils of *Rarh* is limited to eastern part of Chota Nagpur Plateau fringe, covering an approximate area of 7,700 km^2 (comprising the districts of Murshidabad, Birbhum, Bardhaman, Bankura, Purulia, and West Medinipur). A parallel west–east flowing (due to general west to east trending slope of underlying structure) peninsular drainage system (viz., Brahmani, Dwarka, Mayurakshi, Ajay, Damodar, Dwarkeswar, Silai, Kangsabati, and Subarnarekha rivers) dissect the lateritic *Rarh* region into patches of badlands and tropical deciduous forests of West Bengal (an area of 7,700 km^2). In the Rajmahal Basalt Traps, the high elevation zone (200–250 m) is observed in the western part where the basalts are overlain on the Gondwana rock beds. The alluvium zone and the laterites are mostly found in the elevation zone of 0–50 m and 50–100 m, respectively.

The Bengal Basin in its western and south-western parts consists of an easterly inclined shelf, separated from the Precambrian Shield and Gondwana rocks by a prominent zone of dislocation. This region (Fig. 3.2) shows the occurrences of laterite and lateritic soils, the protoliths of which include ferruginous sandstone, red shale, grit, and gravel beds containing dicotyledonous fossil woods (Ghosh and Guchhait 2019). The lateritic and/or ferrisol profile in the laterite plains is 10–20 m thick. Double profiles of laterite have been recorded on Quaternary alluvium that contains abundant lateritic pisolites; these are derived from the adjacent provenance area capped by an in situ laterite profile (Acharyya et al. 2000). It is revealed that the lateritic soils (autochthonous and allochthonous types) of this region are the oldest soils (350–1000 ka, i.e. Early to Middle Pleistocene) found in the Indo-Gangetic Plains (Niyogi 1975; Singh et al. 1998).

The cradle of the world largest Ganga–Brahmaputra Delta is the Bengal Basin which is a structural depression (in between Indian Plate and Eurasian Plate) that the rivers filled up during the last ~150 Ma to its present configuration (Alam et al. 2003; Bandyopadhyay 2007). Fringing the basalts of Early Cretaceous Rajmahal Basalt Traps, sandstones of Gondwana Formations (Permian–Silurian) and granite–gneiss of Archaean Formation at west and Late Quaternary Older Alluvium (Panskura and Sijua Formations) at east, the lateritic zone of *Rarh Bengal* is found at the northwestern border of Bengal Basin. At the west, the Chota Nagpur Plateau has thick regolith envelop of laterites at topographically high level and towards east the laterites are observed as low-level sedimentary units, found in the river valleys and interfluves region. In this region, Niyogi et al. (1970) recognized the oldest terraces of shelf zone of Bengal Basin as ferruginous Lalgarh Formation. As a whole, Table 3.1 gives a short picture of the geological succession in the western shelf zone of the Bengal Basin.

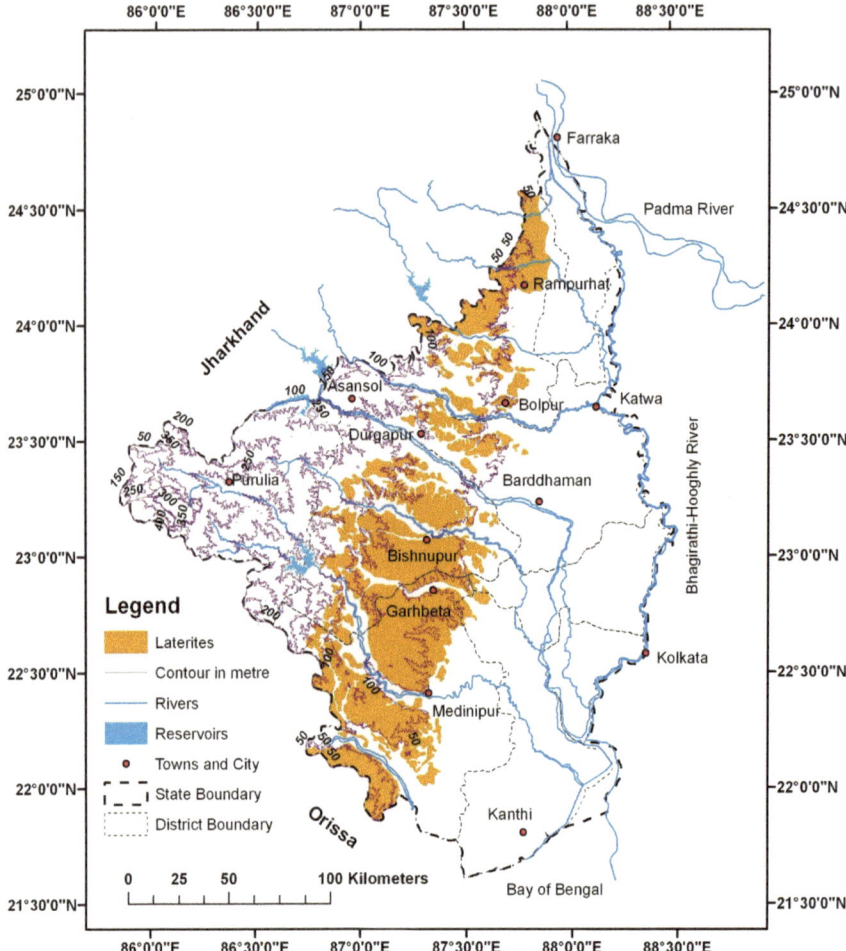

Fig. 3.2 Spatial distribution of laterites (bounded in between 40 and 100 m contour) in western part of the Bengal Basin (Ghosh and Guchhait 2015)

3.2 Bengal Basin

3.2.1 Geological Structure and Tectonics

The different geological dimensions of the Bengal Basin are exceptionally studied by Sengupta (1966, 1972), Alam et al. (2003), Goodbred and Kuehl (2000), Das Gupta and Mukherjee (2006), Bandyopadhyay (2007), Sarkar et al. (2009), Rajaguru et al. (2011), Sinha and Tandon (2014), Akter et al. (2015), Roy and Chatterjee (2015), Valdiya (2016, 2017), and Rudra (2018). The Bengal Basin is a structural and

Table 3.1 Geological succession since Archaean in western part of the Bengal Basin

Lithology	Geological unit	Age
Alternating layers of sand, slit, and clay	Kandi formation	Middle to Late Holocene
Hard clays impregnated with caliche nodules	Rampurhat formation	Late Pleistocene to Early Holocene
Laterite	–	Cainozoic
Rajmahal Basalt Traps	–	Jurassic to Cretaceous
Sandstone and shale	Dubrajpur formation	Triassic to Jurassic
Sandstone and shale with coal seam	Raniganj formation	Upper Permian
Black and grey shale with ironstone, sandstone	Barren measure formation	Middle Permian
Siltstone, sandstone, and Shale with coal seam	Barakar formation	Lower Permian
Pegmatite (unclassified)		Proterozoic
Granite Gneiss	Chota Nagpur Gneissic Complex	Archaean to Proterozoic
Gabbro	Unclassified metamorphics	Archaean to Proterozoic
Quartzite	Unclassified metamorphics	Archaean to Proterozoic
Amphibolite, Hornblende, Schist	Unclassified metamorphics	Archaean to Proterozoic

Source Baksi et al. (1987)

synclinal depression in between two major plates—(1) Indian Plate and (2) Eurasian Plate. The Basin constitutes an enormous area of post-cretaceous sedimentation and tectonic movements extending from about 25° N latitude to about 7° S (Das Gupta and Mukherjee 2006). Most of the area is occupied by the present-day Bay of Bengal, with filled up head of the Bay to the north, continental shelf sediments and coastal sediments of western part and the Assam–Arakan folded mountain belt to the east.

The north-western part of the basin ranges from north of the latitude 18° N and the west of the longitude 90° E. The Indian Peninsular mass has been formed by the fusing together of a number of separate cratons during different Plate Tectonic cycles. The Bengal Basin borders with the peninsular landmass at wet forming Precambrian Gneiss and Schists, Gondwana Supergroup, Early–Late Cretaceous Basalts, Tertiary clays, sandstones and mudstones, Early Pleistocene Laterites and Late Pleistocene to Recent alluvium. The origin of the Basin dates back to the Early Cretaceous times (145 million years BP), when the Gondwanaland was broken up into pieces of plates. Due to northward drift of Indian Plate, the Bengal Basin was formed by the subduction of the Eurasian Plate beneath the Burma Plate (Das Gupta and Mukherjee 2006). There was a soft collision (59–44 million years BP) and recurrent hard collision in the Early Eocene era (44 million years BP), which initiated the Himalayan Orogeny. Thus, a symmetrical pericratonic rift basin was formed in the north-eastern part of

Indian Plate (Alam et al. 2003; Rudra 2018). It is found that a series of en-echelon faults along the western edge of the Bengal Basin are probably related to deep-seated shearing movement in the basement (Das Gupta and Mukherjee 2006).

The western Archaean shield, i.e. Peninsular Craton, which stands along the western border of the Bengal Basin plunges approximately at 87° E meridian and extends further east under thick layers of recent sediments. The basement, on which sediment layers have been deposited, is tilted towards east. The central part of the basin in Bangladesh records nearly 22,000 m thick Early Cretaceous–Holocene sedimentary successions. The hinge zone, a striking structural feature of the Basin (about 25 km width), divides continental and oceanic parts of the Indian Plate and it can be marked by a line connecting Kolkata and Mymensingh (Bangladesh). The N-S trending basin margin fault system (i.e. Chota Nagpur Foothill Fault, Garhmayna–Khandaghosh Fault and Pingla Fault, etc.) deepened the depth of basement from 1–4 km (north) to 5–10 km (South) (Nath et al. 2010; Ghosh and Guchhait 2015).

In relation to western shelf of the Bengal Basin, the Gondwana outcrops of Damodar Graben and Gondwana rocks of Rajmahal along the Chota Nagpur Gneiss Shillong Plateau alignment constitute the limit of western fringe of Basin. Across shelf zone of the Bengal Basin, it has encountered a more substantial development of Gondwanas in a 30 km long east–west stretch (Raniganj Coal Field) at depths ranging from 650 to 3600 m. The gravity modelling confirms that the RBT is well characterized by an elongated nature of relative high Bouguer anomaly (0–25 mGal) along 87° E in contrast to predominantly low Bouguer anomaly (0–40 mGal) in its surroundings of Bengal Basin. This gravity modelling confirms that the Gondwanas (including coals) are preserved in a down-faulted shield edge of the Bengal Basin over an irregular basin floor. In the Bengal Basin, the thickness of sediments overlying the RBT in the wells of West Bengal at Bardhaman, Galsi and Jalangi is respectively 2,515 m, 1,273 m, and 3,216 m.

Neotectonics seems to have directly and indirectly affected the development of landforms and soils in this region. The west–central part of West Bengal (i.e. western shelf zone of the Bengal Basin) is neo-tectonically influenced by the basin margin faults systems (i.e. Chota Nagpur Foothill Fault, Garhmayna–Khandaghosh Fault and Pingla Fault) and the zone of laterites is placed in between the fault systems as an uplifted block (Figs. 3.3 and 3.4). Due to reactivation of some basement faults and tectonic subsidence, eastern and western subunits of the tectonic shelf in the western region were sites of marine transgression during Early Pleistocene and at about 7 ka (Singh et al 1998). In the Holocene uplift of these subunits triggered regression and determined the degree and time of pedogenesis in different units From the presence of a number of important riverine alignments some of these faults may be mildly active even to this day. West of this alignment (which runs a little to the east of the longitude 88° E), the shelf is occupied by a wide NNE–SSW gravity low (Das Gupta and Mukherjee 2006).

The striking feature after Gondwana Supergroup, in the western shelf, is the Rajmahal Basalt Traps which is a thick outcrop of horizontal to sub-horizontal basaltic lava flow of Early Cretaceous time, covering more than 4000 km^2 (Sengupta

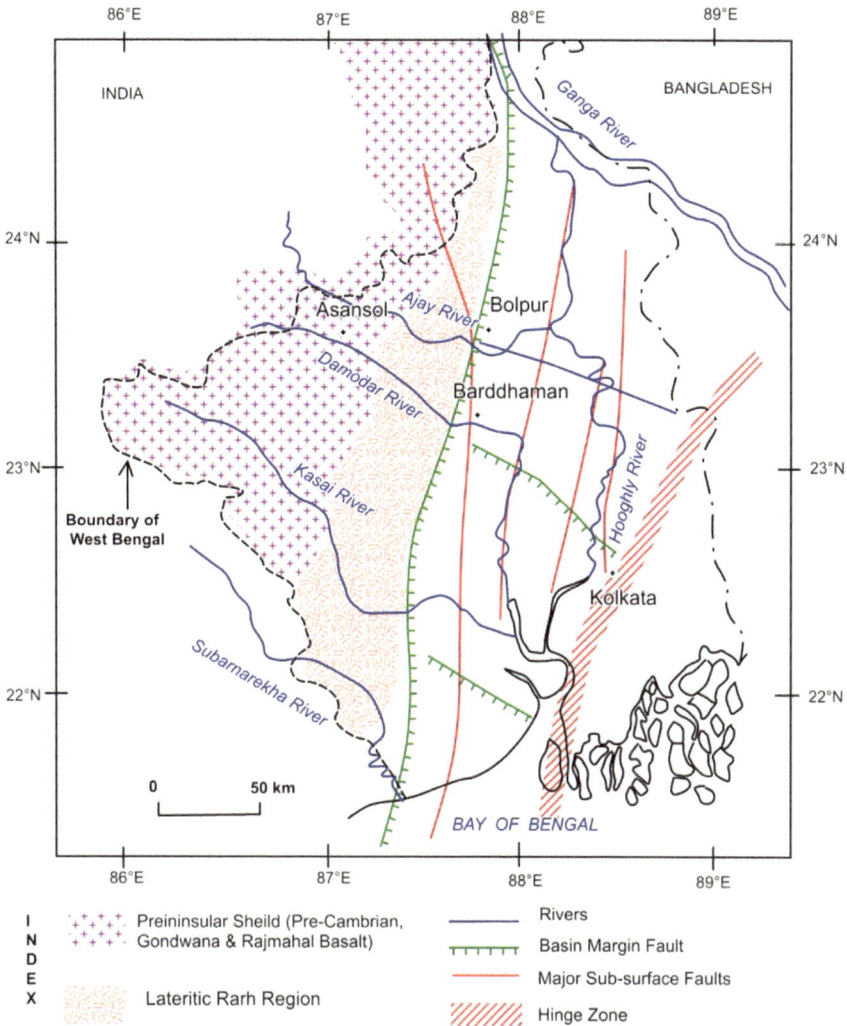

Fig. 3.3 Seismo-tectonic map of the Bengal Basin, showing location of faults and laterite exposures (modified from Das Gupta and Mukherjee 2006)

1972). The Rajmahal effusions are linked to India's passage over the Kerguelen and Crozet Hot Spots, presently in the Indian Ocean and it was generated by decompressional melting of asthenosphere welling up passively beneath the rifted margin of eastern peninsular shield (Sengupta 1972; Baksi 1995; Kent et al. 2002; Mahadevan 2002; Das Gupta and Mukherjee 2006). Though up to 28 flows, aggregating to a thickness of over 331 m have been reported from a borehole drilled in the western shelf of Bengal Basin by the Geological Survey of India (GSI), only 15 flows are demarcated amidst the outcrops of the Rajmahal hills (Mahadevan 2002).

(a)

(b)

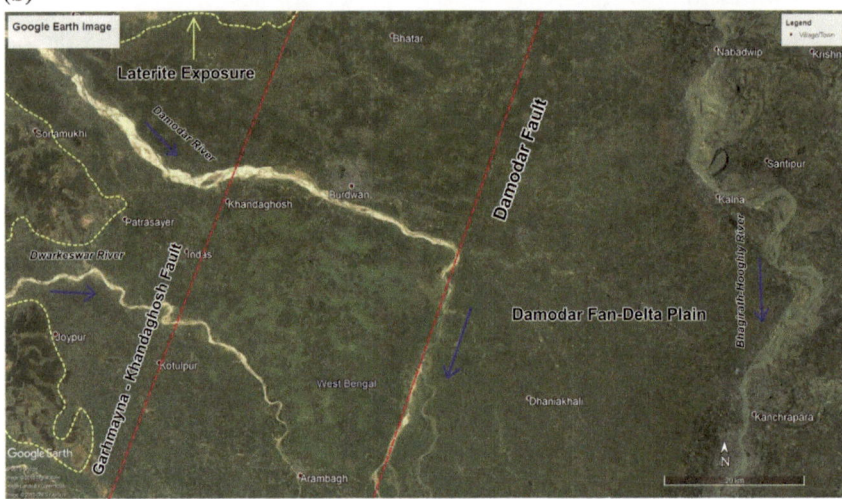

Fig. 3.4 **a** Development of annular drainage pattern due to domal uplift in between two basin margin faults of the Bengal Basin, and **b** deflection of river course due to basin margin fault in the Damodar River (*Source* Google Earth Imagery 2018)

3.2.2 Geomorphic Characteristics

The post-Pilocene sedimentation of shelf zone was brought by the Peninsular Rivers (i.e. Damodar, Ajay, Mayurakshi, forming here sets of terraces along the Basin Margin Fault. The vast alluvial plain higher lateritic tracts and the interlacing channels of the Ganga–Brahmaputra and the Meghna system have given the Bengal Basin a special geomorphic identity. The Bengal Basin is the cradle of the Ganga–Brahmaputra–Meghna (GBM) Delta and characterized by many palaeochannels and shallow swamps (Bandyopadhyay 2007; Rudra 2018). The extensive research shows that the GBM delta is best characterized as a composite system, with different geomorphic regions having morphologic and stratigraphic attributes of an upland fluvial form delta; a lowland and backwater–reach delta; a dwondrift tidal delta and an offshore subaqueous–delta clinoform (Kuehl et al. 2005; Wilson and Goodbred 2015). Located as it is in a tectonically active setting, the GBM delta grew under control of tectonics, high marine and fluvial energy and high influx of Himalayan sediments (Valdiya 2016). About 10^2 m^3 of water with 10^9 tonnes of sediment per year make this system morphologically active (Akter et al. 2015). In the last five decades, the GBM delta has prograded at a rate of 17 km^2 per year, whereas most large deltas elsewhere in the world suffer from sediment starvation (Wilson and Goodbred 2015; Akter et al. 2015).

The lateritic uplands on the western margin of the basin are followed successively by palaeodelta plain, younger fluviodelta plain and recent tidal flat of Sundarbans with beach ridges (Valdiya 2017). The first phase of sediment filling was initiated by the western tributaries to the Bhagirathi–Hooghly River. The rivers like the Mayurakshi, the Ajay, the Damodar, and the Rupnarayan draining out of the western uplands formed palaeo-/subdeltas at the their respective outfalls along the oldest strandline which ran north-eastwards from Digha to Nabadwip (Agarwal and Mitra 1991; Rudra 2018). Dendritic valleys cuts that are filled with younger flood-delta plain sediment penetrates as fingers with older alluvial plain (Acharyya et al. 2000). Likewise, valley-cut fingers with older alluvium penetrate into the lateritic plain. The nature of the palaeochannels and alluvium from the older alluvium and lateritic plain are atypical of deltas. In Anthropocene, the natural decay of rivers has been exacerbated by the human intervention, especially where rivers are embanked and no allowance made for their migration through meandering and avulsion.

Goodbred and Kuehl (2000) identified two major stratigraphic facies in the Late Quaternary sequence of southern West Bengal—oxidized facies and different facies of sands. The oxidized facies consists of stiff slity clays brown to orange in colour and is distributed locally through out the region at a depth of 10–45 m underlying this mud up to 10 m of weathered, iron-stained, heavy mineral–deficient sands are also part of the oxidized facies. That unit was deposited 14,000 years BP. The GBM delta took the present shape in 3000 years BP and the post-Pleistocene rise of the sea level facilitated the formation of the Holocene delta. Stanley and Warne (1994) opined that the GBM delta started to grow since 7060 + 120 years BP, but Goodbred and Kuehl (2000) fixed the time 10,000–11,000 years BP for the evolution of the GBM delta. Sedimentology, carbon isotope and sequence stratigraphic analysis of subsurface sediments from the GBM delta plain shows that during the Last Glacial Maximum (LGM) sea-level lowering of >100 m produced a regional unconformity, represented by ferruginous and calcareous palaeosols and incised valley in the western shelf zone (Sarkar et al. 2009). Intensification of monsoon and very high sediment discharge (~4–8 times than modern) caused a rapid aggradation of both floodplain and estuarine valley fill deposits between 8 and 7 ka (Sarkar et al. 2009; Sinha and Tandon 2014).

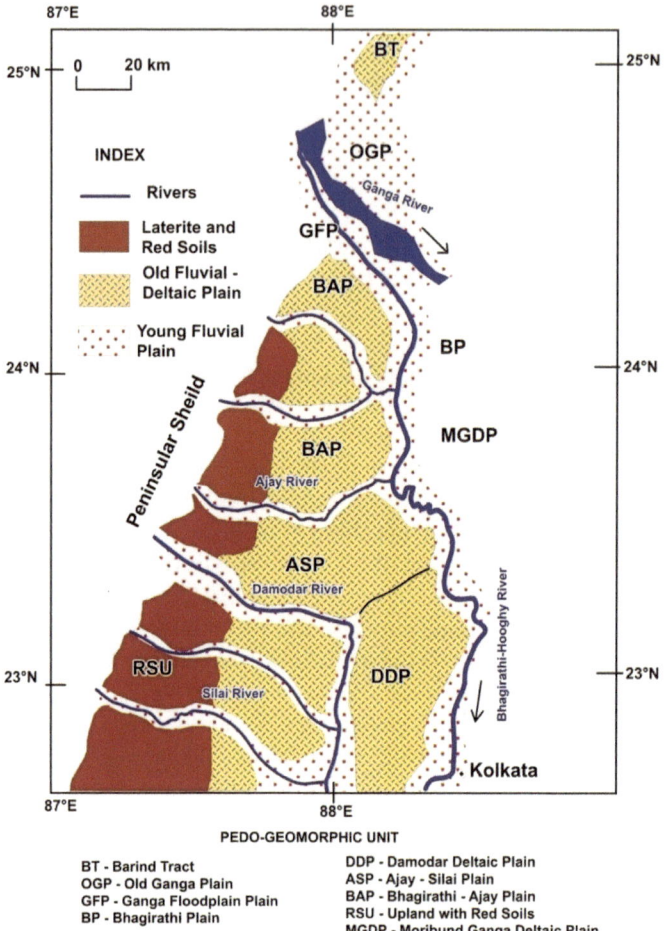

PEDO-GEOMORPHIC UNIT

BT - Barind Tract	DDP - Damodar Deltaic Plain
OGP - Old Ganga Plain	ASP - Ajay - Silai Plain
GFP - Ganga Floodplain Plain	BAP - Bhagirathi - Ajay Plain
BP - Bhagirathi Plain	RSU - Upland with Red Soils
	MGDP - Moribund Ganga Deltaic Plain

Fig. 3.5 Important pedo-geomorphic units of the western part of GBM delta (modified from Singh et al. 1998)

The OSL dating data from the recent sediments of GBM delta shows that peat facies developed around 6.9 ka BP which today occurs 2 m below the present sea level. The overlying 7 m thick silty-clayey sediments are non-marine alluvial origin and suggest progradation of delta after 6.9 ka BP (Rajaguru et al. 2011). In Holocene subsequent uplift, pedogenesis and flood sedimentation developed six pedo-geomorphic units in the western shelf zone following the oldest lateritic *Rarh* Plain at west (Middle–Late Pleistocene) (Fig. 3.5). OSL (Optically Stimulated Luminescence) dating ages estimates obtained from basal samples for the soils of different pedo-geomorphic units of the GBM delta are: (i) Ganga Floodplain—0.5 ka, (ii) Old Ganga Plain—1.5 ka, (iii) Bhagirathi Plain—1 ka, (iv) Damodar Deltaic Plain—3.6 ka, (V) Ajay-Silai Plain—5.4 ka, and (vi) Bhagirathi–Ajay Plain—6.7 ka (Singh et al. 1998).

The undulating plateau fringe along with the adjoining lateritic plains lying in between the Chota Nagpur Plateau and the Bengal Basin is described as 'Rarh Plain'. The elevation ranges of the region from 50 to 120 m, having average slope of 3°–4° towards east and south-east. The region of red latosol is occupied by dry deciduous forest, mainly *Sal*, and due to excessive rill and gully erosion the many patches of badlands are developed in this lateritic terrain. The identification of *Rarh* Plain within the western shelf zone of Bengal Basin and the genesis of different laterites have been debated long. This book tries to unearth the characterization and geomorphic evolution of lateritic *Rarh* Plain through scientific approaches of Regolith Geology and Geomorphology.

3.3 Climate

The climate of this region has been identified as sub-humid and subtropical monsoon type, receiving mean annual rainfall of 1300–1420 mm. About 75% of annual rainfall is received from south-west monsoon from June to September. The mean annual air temperature is 26.2° C. Mean summer air temperature (April, May, and June) is 35° C and mean winter air temperature (December, January, and February) is 19° C. It has been found that mean summer soil temperature is 31.6° C and mean winter soil temperature is 24.1° C respectively. The soil temperature regime is 'hyperthermic', and soil moisture regime is 'ustic'. The many parts of region are affected by prolonged heavy rainfall during tropical cyclone or depression and floods occur in the lower part of river basins. The annual potential evaporationtranspiration (PET) varies from 1400 to 1600 mm. Highest value of PET is observed in the month of May and it is lowest in the month of December and January. The moist period usually starts in May and humid period starts in June and ends almost in the second week of October (Table 3.2).

The peak monsoon and cyclonic rainfall intensity of 21.51 mm h^{-1} (minimum) to 25.51 mm h^{-1} (maximum) is the most powerful climate factor to develop this lateritic badlands. The recorded maximum and minim temperature is 45° C (April–May) and 9° C (December–January) respectively, with seasonal variation of 15–19° C. The period between mid-June and September is the active erosion phase due to heavy downpours, removing ferruginous sediments from the gullied catchments (Ghosh and Bhattacharya 2012). The region is experiencing intense thunderstorms during hot summer and prolonged rainfall during the tropical depression and cyclone. The present climate of this region reflects a unique morpho-genetic region—Tropical Wet-Dry Zone (Koppen A_w Climate) of Planation Surface Formation (Chorely et al. 1984) where chemical weathering, sheet floods, and hillslope erosion are dominant with the development of red loam kaolin-rich ferruginous materials, tors and badlands are mostly developed. The identified dominant lateritization processes are cementation and induration of Fe–Al mottles (haematite and goethite) by centripetal accumulation and secondary accumulation of kaolinite forming rounded to elliptical-shaped nodules and meta-nodules, vertical leaching of silica and bleaching of saprolite.

Table 3.2 Climatological data of study area (Siuri, Birbhum district)

Month	Temperature (°C)			Relative humidity (%)	Rainfall (mm)
	Daily maximum	Daily minimum	Average		
January	25.2	12.0	18.6	57	15.1
February	28.3	14.4	21.3	49	15.9
March	34.0	19.3	26.6	43	21.5
April	38.2	23.2	30.7	50	33.6
May	38.4	25.1	31.7	62	71.9
June	35.4	25.6	30.5	75	212.4
July	32.5	25.1	28.8	82	289.0
August	32.0	25.1	28.5	83	305.2
September	31.9	24.6	28.2	81	284.0
October	31.2	22.4	26.8	74	118.5
November	28.9	17.1	23.0	63	12.2
December	25.9	12.7	19.3	58	41.1

Source Sarkar et al. (2007)

3.4 Natural Vegetation and Land Use

The vegetations of the district belong to the tropical dry type deciduous with few evergreen species occurring here and there. The tree species occurring in the forests of undulating lateritic uplands along the *Rarh* Plain comprise of *Sal* (*Shorea robusta*), *Haldu* (*Adina cordifolia*), *Pial* (*Anogeissus lattfolia*), *Mahua* (*Madhua indica*) and *Peasal* (*Palerocarpus marsupium*). The other important tree species are *Arjun* (*Terminalia arjuna*), *Sirish* (*Albizzia lebbeck*), *Bakul* (*Mimusops elengi*), *Simul* (*Bombax ceiba*), *Kend* (*Diospyros melanoxoilon*), and *Palas* (*Buiea monosperma*). In the direr part, the characteristics shrubs and herbs include *Acacia Bridelia, Calotropis, Capparis, Casia, Streblus* and *Feronia* etc. (Ghosh and Guchhait 2019). Though once upon a time the most of the region was covered under thick forest, mainly Sal; due to encroachment of stone crushers, mining and agriculture the forests are fragmented and vanished from some places. The agricultural productivity is very low and most of the arable land remains fallow in dry season. The main crops are paddy, maize, corn, and oilseed. Another distinctive land use of this region is stone mining and *morrum* mining which have modified the slope and morphology of the basaltic hills and the lateritic lands. Gradually it engulfs the afforested region (Acacia Plantation). In this area slope and topography has been modified and natural forest cover is also destroyed, but now acacia plantation covers the region. In the grassland parts, livestock grazing is observed and most of the land is infertile barren lateritic cover which has surface crusting of ferruginous materials (Fig. 3.6).

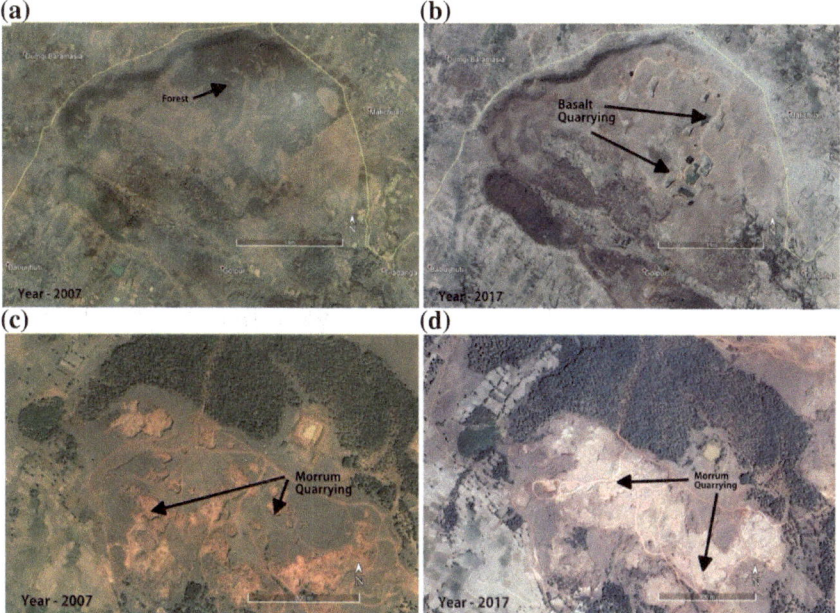

Fig. 3.6 Identifying the Changes in land use through Google Earth Imagery—**a** and **b** the forest land are vanished due to development basalt quarrying in between 2007 and 2017, **c** and **d** the lateritic land is engulfed by the morum quarrying in between 2007 and 2017

3.5 Pedological Features of Soils

Most of the soils over laterites, granite–gneiss and basalts are in group of latosol. These soils with varying depth (5–20 cm) are shallow to moderately shallow, well drained and highly ferruginized and occur on gently to moderately sloping plateau fringe. Few soils are loamy–skeletal with 35–40% gravels in the soils series. Generally soils have low available water capacity and underlain by laterites with weathered rock fragments and quartz gravels. These soils are shallow, light-textured and gravely with very low moisture retention capacity. The soils are susceptible to severe erosion and should be under vegetative cover and afforestation.

In this region, fours distinct soil series are recognized in four sample sections of *Rarh* Plain—(1) Type A (Rampurhat—24° 11′ 44″ N, 87° 44′ 03″ E), (2) Type B (Rangamati—22° 24′ 42″ N, 87° 17′ 55″ E), (3) Type 3 (Panagarh—23° 27′ 10″ N, 87° 31′ 51″ E), and (4) Type 4 (Bolpur—23° 40′ 18″ N, 87° 39′ 10″ E) (Table 3.3). In short, the dark reddish to brown-coloured sandy clay loam of 0–20 cm (A horizon, maximum grass-root zone) is developed over the fragmented secondary laterites. These soil series have weak fine crumb and granular structure (slightly hard, friable, and slightly sticky), 2–5 mm size of manganese nodules, >2 mm size of ferruginous nodules with goethite cortex, 30–40% gravels and pebbles, excessive drained surface

Table 3.3 Short description of soil series found in the study area

Layers	Depth (cm)	Pedological descriptions
1. Soil series—Type A		
A	0–16	Dark reddish brown sandy clay loam; weak fine crumb and granular stricture; slightly hard, friable, slightly sticky, and slightly plastic, 30–40% gravels; medium and fine pores; pH 5.4, abrupt wavy boundary
C_r	16–34	Weathered rocks with ferruginous concretions
2. Soil series—Type B		
A_1	0–12	Strong brown and brown to dark brown loamy sand; weak medium crumb structure; slightly hard, friable pores and insect channels; very few hard Fe–Mn concretions; pH 6.0; clear wavy boundary
B_w	12–31	Strong brown and brown to dark brown gravely sandy loam; weak, medium sub-angular blocky structure; slightly hard, friable, and slightly sticky; many fine and coarse roots; many fine and coarse gravels, pH 6.2; gradual wavy boundary
C_r	31–52	Yellowish red to dark reddish brown; laterite mass with quartz fragments
3. Soil series—Type C		
A_1	0–9	Strong brown sandy clay loam; weak medium sub-angular blocky structure; slightly hard, friable, slightly sticky, and slightly plastic; many fine and medium roots; appreciable amount of gravels; pH 6.0; clear smooth boundary
B_{w1}	9–35	Strong brown sandy clay loam; weak, fine sub-angular blocky structure; slightly hard and plastic; few fine roots;, appreciable gravels; pH 6.2; clear smooth boundary
B_{w2}	35–42	Yellowish red sandy clay loam; weak, fine sub-angular blocky structure; slightly hard, friable, and sticky; few fine roots; appreciable gravels; Fe–Mn and Fe–AL mottles; pH 5.8; clear smooth boundary
C_r	42–56	Weathered granite–gneiss and quartz fragments

and pH of 5.4–5.7 (Table 3.4). The loose secondary laterite (20–44 cm) is developed as cementation (low cohesion and weak structure) of derived materials over mottle and kaolinite horizon and it is much prone to overland flow erosion, tunnel erosion, and bank failure. The percentage of coarse and find sand varies from 49.1 to 70.2% in these soil series, whereas the percentage of silt varies from 19.1 to 28.3% and clay ranges in between 10.7 and 29.1%. Down the slope or at the base of hillslope the thickness of soils and secondary laterites is much more than convex part. The percentage of coarse materials (>74%) is found at the top of slope and the percentage of silt and clay (>45%) is found at the toe of slope.

Table 3.4 Analytical laboratory-based soil data of four series

| Soil series | Horizon | Depth (cm) | Particle size diameter (mm) | | | Coarse fragments > 2 mm % whole soil | Organic carbon % | pH |
			Sand (2.0–0.05)	Silt (0.05–0.002)	Clay (<0.002)			
Type A	A	0–16	49.1	28.3	22.6	38.0	1.3	5.4
	C_r	16–30	Lateritized regolith					
Type B	A	0–12	81.6	7.3	11.1	5.0	0.57	6.0
	B_w	12–31	73.5	8.4	18.1	40.0	0.60	6.2
	C_r	>31	Lateritized regolith					
Type C	A	0–9	54.0	13.6	32.4	23	0.68	6.0
	B_{w1}	9–35	72.0	7.0	21.0	25	0.28	6.2
	B_{w2}	35–42	66.0	9.9	24.1	33	0.18	5.8
	C_r	42–56	Lateritized regolith					
Type D	A	0–12	84.0	4.4	11.6	Nil	0.21	5.2
	B_{wt1}	12–32	70.6	10.8	18.6	5	0.21	4.9
	B_{wt2}	32–58	67.2	10.2	22.6	20	0.20	5.1
	C_r	58–73	Lateritized regolith					

3.5.1 Type A Soil Series

This series of soil has been classified as loamy–skeletal and hyperthermic Lithic Ustorthents, having texture of sandy loam and dark reddish brown colour. The soils are excessively drained and oxidized (crust formation in upper layer), creating barren wasteland with sparse bushy vegetation. The soil is evolved in moderately sloping undulating plateau of basalt, granite, and gneiss (70–150 m elevation range). The dark reddish brown A horizon (0–16 cm) is sandy clay loam, having weak fine, crumb, and granular structure and 2–5 mm size manganese nodules with 30–40% gravels. The parent material C_r horizon (16–34 cm) is the weathered rocks and regolith with ferruginous concretions. The pH of soil horizons varies from 5.4 to 5.6.

3.5.2 Type B Soil Series

This soil series has been classified as loamy-skeletal and sandy loam texture. The soil is formed in gently sloping lateritic plateau fringe (60–70 m elevation range). Strong brown to dark brown-coloured A horizon (0–12 cm) has weak medium crumb structure, friable and few hard iron–manganese concretions, having only 5% of coarse fragments, but gravelly sandy loam B_w horizon (12–31 cm) has weak sub-angular blocky structure and many fine and coarse gravels (40%). The C_r horizon (31–52 cm) is yellowish red to dark reddish brown laterite mass. The pH of soil horizons varies from 6.0 to 6.2.

3.5.3 Type C Soil Series

The soil is categorized fine loamy hyperthermic and sandy clay loam texture group.. The soil is formed in gently sloping terrain of plateau fringe (80–90 m elevation range). Strong brown-coloured A horizon (0–9 cm) has weak medium sub-angular blocky structure, slightly hard–friable, having 23% of coarse fragments. Both B_{w1} and B_{w2} horizons have increasing amount of sand content (66.0–72.0%), showing high degree of leaching. C_r horizon (>42 cm) is ferruginous regolith having ample amount of quartz. The pH of soil horizons varies from 5.8 to 6.0.

3.5.4 Type D Soil Series

The fine loamy and hyperthermic sandy clay loam texture soil is formed on the gently sloping undulating terrain (50–60 m elevation range). A horizon (0–12 cm) includes reddish yellow and brown sandy loam, weak fine sub-angular blocky structure, having only 11.6% of clay. B_{t1} horizon (12–32 cm) is characterized by strong brown sandy loam, patchy thin clay coatings on ped faces, and little semi-hard iron–manganese concretions. B_{t2} (32–58 cm) is yellowish red sandy clay loam, having weak medium sun angular blocky structure, clay coatings and 20% of coarse fragments. The parent material C_r is lateritic mass. The pH of soil horizons varies from 4.9 to 5.2.

References

Acharyya SK, Lahiri S, Raymahashay BC, Bhowmik A (2000) Arsenic toxicity of groundwater in parts of the Bengal Basin in India and Bangladesh: the role of Quaternary Stratigraphy and Holocene sea-level fluctuation. Environ Geol 10:1127–1137

Agarwal RP, Mitra DS (1991) Palaeogeographic reconstruction of Bengal Basin during Quaternary period. In: Vaidyanadhan R (ed) Quaternary deltas of India. Memoir of the Geological Society of India, Bangalore, pp 13–24

Akter J, Sarker MH, Popescu I, Roelvink D (2015) Evolution of the Bengal Delta and its prevailing processes. J Coast Res 32(5):1212–1226

Alam M, Alam MM, Curray JR, Chowdhary MLR, Gandhi MR (2003) An overview of the sediment geology of the Bengal Basin in relation to the regional tectonic framework and basin-fill history. Sediment Geol 155(3–4):179–208

Bagchi K, Mukherjee KN (1983) Diagnostic survey of Rarh Bengal (Part II). University of Calcutta, Calcutta

Baksi AK (1995) Petrogenesis and timing of volcanism in the Rajmahal flood basalt province, north-eastern India. Chem Geol 121:73–90

Baksi AK, Barman TR, Paul DK, Farman E (1987) Widespread early cretaceous flood basalt volcanism in eastern India: geochemical data from the Rajmahal-Bengal-Sylhet Traps. Chem Geol 63:133–141

Bandyopadhyay S (2007) Evolution of the Ganga–Brahmaputra Delta: a review. Geogr Rev India 69(3):235–268

Biswas A (1987) Laterities and lateritoids of Bengal. In: Datye VS, Diddee J, Jog SR, Patial C (eds) Exploration in the tropics. K.R. Dikshit Felicatiobn Committee, Pune, pp 157–167

Bourman RP (1993) Perennial problems in the study of laterite: a review. Aust J Earth Sci 40(4):387–401

Das Gupta AB, Mukherjee B (2006) Geology of N.W. Bengal Basin. Geological Society of India, Bangalore

Gajbhiye K S (2007) Optimizing land use of Birbhum District (West Bengal) soil resource assessment. National bureau of soil survey and land use planning, NBSS Publ., Nagpur

Ghosh S, Bhattacharya K (2012) Multivariate erosion risk assessment of lateritic badlands of Birbhum (West Bengal, India), a case study. J Earth Syst Sci 121(6):1441–1454

Ghosh S, Guchhait S (2015) Chraterization and evolution of primary and secondary laterites in northwestern Bengal Basin, West Bengal, India. J Palaeogeogr 4(2):203–230

Ghosh S, Guchhait SK (2019) Modes of formation, Palaeogene to Early Quaternary Palaeogenesis and geochronology of laterites in Rajmahal Basalt Traps and Rarh Bengal of Lower Ganga Basin. In: Das BC, Ghosh S, Islam A (eds) Quaternary geomorphology in India. Springer, Singapore, pp 25–60

Goodbred SL, Kuehl SA (2000) Enormous Ganges–Brahmaputra sediment discharge during strengthened Early Holocene monsoon. Sediment Geol 28:1083–1080

Kent RW, Pringle MS, Mullar RD, Saunders AP, Ghose NC (2002) [40]Ar/[39]Ar geochronology of the Rajmahal Basalt, India and their relationship to the Kerguelen Plateau. J Petrol 43(7):1141–1153

Kuehl SA, Allison MA, Goodbred SL, Kudrass H (2005) The Ganges–Brahmaputra Delta. In: Giosan L, Bhattacharya J (eds) River deltas—concepts, models and examples, vol 83. SEPM Society for Sedimentary Geology Special Publication, pp 413–434

Mahadevan TM (2002) Geology of Bihar and Jharkhand. Geological society of India, Bangalore

McFarlane MJ (1976) Laterite and landscape. Academic Press, London

Nath SK, Thingbaijam KKS, Vyas JC, Sengupta P, Dev SMSP (2010) Macroseismic-driven site effects in the southern territory of West Bengal, India. Seismol Res Lett 81(3):480–487

Niyogi D (1975) Quaternary geology of the coastal plain in West Bengal and Orissa. Indian J Earth Sci 2:51–61

Niyogi D, Mallick S, Sarkar SK (1970) A preliminary study of laterites of West Bengal, India. In: Chatterjee SP, Das Gupta SP (eds) Selected papers physical geography (vol 1). 21st international geographical congress, Calcutta, National Committee for Geography, pp. 443–449

Pascoe EH (1964) A manual of the geology of India and Burma, vol 3. Geological Survey of India, Delhi

Pedro G (1968) Distribution des principaux types d'alteration chimique a la surface du globe. Presentation d'une esquisse geographique. Rev Geogr Phys Geol Dyn 2(10):457–470

Pendleton RL (1936) On the use of the term laterite. Am Soil Surv Bull 17:102-B

Rajaguru SN, Deotare BC, Gangopadhyay K, Sain MK, Panja S (2011) Potential geoarchaeological sites for luminescence dating in the Ganga Bhagirathi-Hugli delta, West Bengal, India. Geochronometria 38(3):282–297

Roy AB, Chatterjee A (2015) Tectonic framework and evolutionary history of the Bengal Basin in the Indian subcontinent. Curr Sci 109(2):271–279

Rudra K (2018) Rivers of the Ganga–Brahmaputra–Meghna Delta. Springer, Singapore

Sarkar D, Dutta D, Nayak DC, Gajbhiye KS (2007) Optimizing land use of Birbhum District (West Bengal) soil resource assessment. National bureau of soil survey and land use planning, NBSS Publ., Nagpur

Sarkar A, Sengupta S, McArthur JM, Ravenscroft P, Beer MK, Bhusan R, Samanta A, Agrawal S (2009) Evolution of Ganges-Brahmaputra western delta plain: clues from sedimentology and carbon isotopes. Quat Sci Rev 28:2564–2581

Sengupta S (1966) Geological and geophysical studies in western part of Bengal Basin, India. AAPG Bulletin 50(5):1001–1017

Sengupta S (1972) Geological framework of the Bhagirathi-Hooghly Basin. In: Bagchi KG (ed) The Bhagirathi-Hooghly Basin: proceedings of the interdisciplinary symposium, Calcutta, pp 3–8

Singh LP, Parkash B, Singhvi AK (1998) Evolution of the lower gangetic plain landforms and soils in West Bengal, India. Catena 33:75–104

Sinha R, Tandon SK (2014) Indus-Ganga-Brahmaputra plains: the alluvial landscape. In: Kale VS (ed) Landscapes and landforms of India. Springer, Dordrecht, pp 53–63

Stanley DJ, Warne AG (1994) Worldwide initiation of Holocene marine deltas by deceleration of sea-level rise. Sci 265(5189):228–231

Tardy Y (1992) Diversity and terminology of laterite profile. In: Martini IP, Chesworth W (eds) Weathering, soils and paleosols. Elsevier, Amsterdam, pp 379–405

Valdiya KS (2016) The making of India, geodynamic evolution. Springer, Dordrecht

Valdiya KS (2017) Ganga-Brahmaputra Plains. In: Valdiya KS, Sanwal J (eds) Neotectonisim in the Indian Subcontinent. Elsevier, New York, pp 151–184

Wilson CA, Goodbred SL (2015) Construction and maintenance of the Ganges–Brahmaputra–Meghna Delta: linking process, morphology and stratigraphy. Ann Rev Mar Sci 7.67–78

Chapter 4
Characterization of Primary Laterite Profiles

Abstract In this chapter, it is focused to understand the effect of tropical weathering on the gneiss, basic dyke, Gondwana group of rocks and Rajmahal Basalt Traps. Merely these primary laterites on different rocks have massive appearance (in situ weathering) reflecting vermicular lateritic crust, mottled zone with lithomarge clay and deeply weathered–altered rocks, called saprolite.

Keywords Duricrust · Pisolitic ferricrete · Murram · Saprolite · Parent rock

4.1 Laterite over Gneiss

The geological formations of granite, garnet–gneiss and pegmatite–gneiss are encountered in some pockets of western Birbhum district, fringing Rajmahal Basalt Traps and Gondwana formation. The field study reveals that development of true laterite profiles are quite absent here and sub-surface chemical weathering does not produce saprolite. In three samples, sites of Bataspur (23° 55′ 47″ N, 87° 29′ 17″ E), Rautrara (23° 53′ 12″ N, 87° 28′ 15″ E), and Sukna (23° 58′ 55″ N, 87° 30′ 52″ E) 1.5–2.2 m thick laterite profiles are found, having 60–80 cm thick duricrust layer (Fig. 4.1). The post-Pleistocene pedogenesis developed 28–55 cm ferralitic latosol over the laterites. The soils are reddish colour, coarse grained, and containing relict parts of duricrust. The 50–65 cm thick pisolitic ferricrete and *murram* unit overlies on gneiss with transitional contact with little weathering attributes. The duricrust unit is pisolitic and rare vermicularity. Though granite–gneiss terrain is a good source of alumina, such alumina rich zones are not developed over gneiss.

4.2 Laterite over Dolerite

The development of laterite profile over dolerite dyke within the Chotanagpur Granite–Gneissic Complex is quite rare in India. In an exposed *murram* quarry at Husennagar (23° 48′ 04″ N, 87° 08′ 34″ E) of Birbhum, a 4.0 m thick laterite profile is

S. Ghosh and S. K. Guchhait, *Laterites of the Bengal Basin*,
SpringerBriefs in Geography, https://doi.org/10.1007/978-3-030-22937-5_4

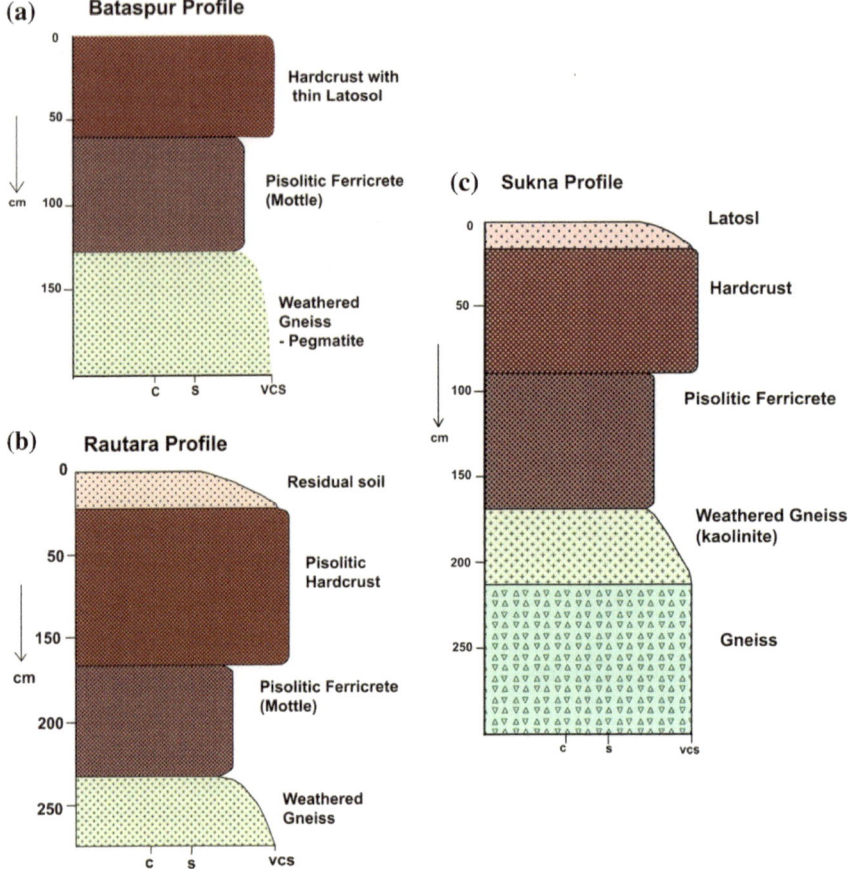

Fig. 4.1 Development of hardcrust, pisolitic ferricrete and coarse saprolite on Chotanagpur Gneiss at **a** Bastapur, **b** Rautara, and **c** Sukna of Birbhum district

analysed to understand the in situ or autochthonous type of development (Fig. 4.2). A 33 cm thick dismantle gritty horizon of Fe nodules overlies on the 1.3 m thick pisolitic ferricrete. About 2.3 m thick saprolite zone is developed here having litho-relicts and kaolinite. In between hardcrust and saprolite zone, a *murram* zone of loose ferricrete, ferruginized feldspar and rock fragment occurs. The pisolitic ferricrete zone consists of mainly iron oxides–hydroxides with goethite coating. This hardcrust zone is not so much hard and compact as it is found in other ferruginized hardcrust. The petrographic study of dolerite reveals that it is constituted of pyroxene, amphibole, plagioclase and opaque as major phases and sphene, epidote and secondary carbonate as accessory phases. The saprolite is the altered rock zone where the texture of relict rock is still preserved. It is reddish green in colour and at places it is altered to whitish clayey materials with glimpses of altered plagioclase and amphibole grains.

Fig. 4.2 Formation of latosol, pisolitic ferricrete, and saprolite on dolerite at Husennagar of Birbhum district

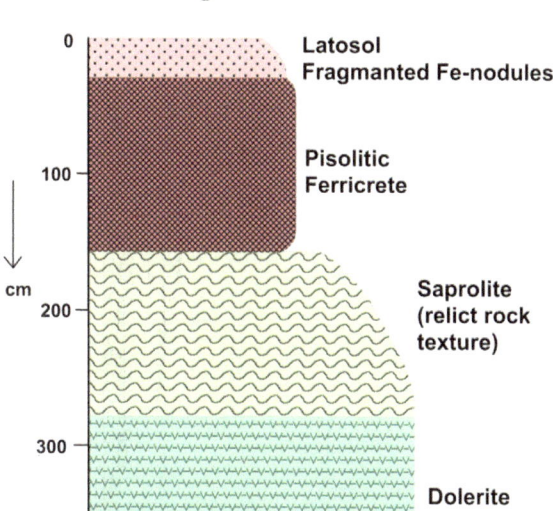

4.3   Laterite over Gondwana

Geomorphologically, the terrain of Gondwana represents hummock type upland area, dissected by numerous tributaries of Ajay and Damodar rivers and the sequences of lithology are characterized by sandstone–clay–fire clay–sandstone. On most cases, the laterites are developed over first unit of sandstone and clay. The detailed studies of lithofacies are conducted in the sample sites at Bastabpur (23° 48′ 50″ N, 87° 13′ 29″ E), Hetampur (23° 47′ 03″ N, 87° 23′ 37″ E), and Deucha (23° 24′ 06″ N, 87° 14′ 55″ E) of Birbhum district and at Barjora (23° 25′ 41″ N, 87° 16′ 32″ E) of Bankura district.

4.3.1   Bastabpur Profile

A section of laterite developed over alternate the sandstone–clay–sandstone unit where the upper sandstone is weathered to form ferruginous hardcrust (Fig. 4.3). A lithosection of about 9.0 m is presented in Fig. 4.3 with characteristic sediments and ferricrete zone. The whole sequence is characterized by 1.5 m thick ferricrete zone followed successively by 20 cm ferruginized very hard pisolitic band, alerted structureless sandstone (with pisolites) of 0.9–1.0 m thick, 0.6 m thick bedded sandstone, massive sandstone (0.65–0.70 m), 35 cm thick clay bed, 0.4 m thick fragmented

Fig. 4.3 Laterite profile
developed over Gondwana
sequence at Bastabpur of
Birbhum district

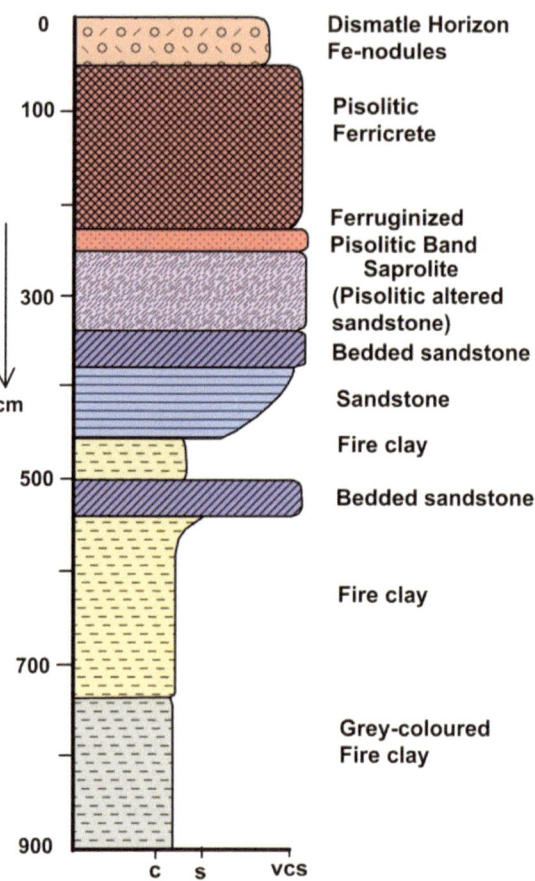

bedded sandstone, and 3.2 m thick fireclay unit. The pisolitic ferricrete zone is characterized by iron pisolites and ferricrete nodules whose core is constituted of sand. The ferruginized pisolitic band occurs as a continuous band parallel of bedding of the sequence. The lower sandstone unit is also affected by some extent of lateritization as evidenced by the presence of some pisolitic nodules within. So in this sequence of Gondwana the extent of lateritization extend up to a depth of maximum 3.5 m.

4.3.2   Hetampur Profile

At a clay quarry, the Gondwana sequence is lateritized to certain extent to form pisolitic ferricrete zone at top, but no such mottle zone has been developed at that

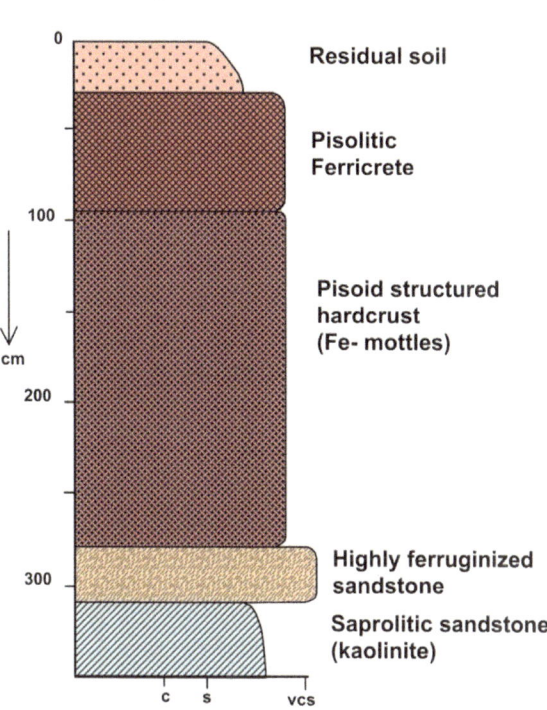

Fig. 4.4 Development of pisolitic ferricrete over Gondwana sandstone at Hetampur of Birbhum district

site (Fig. 4.4). A litho-log profile of 9 m is presented here with characteristic disman-tle gritty zone at top, glaebular zone at middle and saprolite at base. The ferricrete zone of 2.4 m thick is followed successively by 20 cm ferruginized hard pisolitic band. The pisolitic ferricrete zone is characterized by Fe–Al pisolites and ferricrete nodules whose core is constituted of sand. The ferruginized pisolitic band occurs as a continuous band parallel to the bedding of the Gondwana sequence. The lower sandstone is affected by lateritization as evidenced by the presence of some pisolitic nodules within it. The range of lateritization extends up to the depth of maximum 3.5 m in this site. The underlying parent rocks sequences are marked as (i) ferrug-inized sandstone of 0.9–1.0 m thick, (ii) bedded sandstone of 0.6 m thick, (iii) clay bed of 40 cm, (iv) 0.42 m fragmented bedded sandstone, and (v) ash–grey colour fireclay unit of 3.4 m.

4.3.3 Deucha Profile

At this site, a well-developed 3 m primary laterite profile overlies on the Gondwana sandstone which is transformed into ferruginized sandstone and altered saprolite

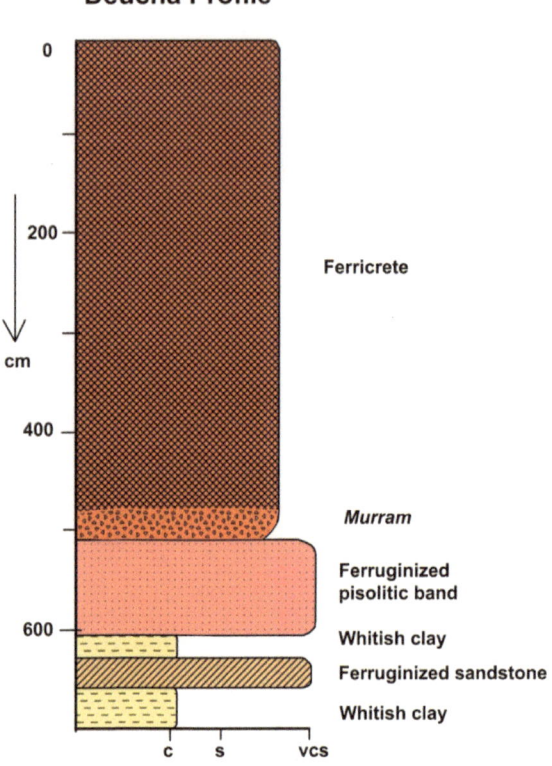

Fig. 4.5 Thick ferricrete and murram development over ferruginized sandstone–clay beds at Deucha of Birbhum district

(Fig. 4.5). The residual latosol of 25–30 cm thick develops over the ferricrete. The presence of loose *murram* over the hardcrust is not a continuous phenomenon but wherever it is present, it occurs as 30–40 cm thick unit constituting of hardcrust fragments, circular to amoeboid shaped concretions of iron and sand balls. Pisolitic hardcrust of 0.8–1.2 m thick is very compact and hard in nature and constituted of globular pisolites whereas the underlying pisoid structured part (2.0–2.2 m thick) is characteristically rich in pisoid and fluid passage paths. These pisoids vary in shape and size with sandy core and brownish iron-rich rim. In the ferricrete zone, the inter-space of pisoids or the channel ways are characterized by kaolin rich clay and gibbsite. The cut surface of the hardcrust exposes branching of tubular channels which are the path of SiO_2 leaching. The ferricrete zone is underlain by very hard sandstone (10–60 cm thick) which is highly ferruginized by lateritization to form altered sandstone or coarse saprolite at base (45–50 cm thick).

4.3.4   Barjora Profile

In the laterite capping hummock 7 m thick laterite profile is found to be overlying the sandstone–clay sequence of Gondwana which is the alteration of sub-arkosic to arkosic sandstone. One of the main laterite sections represented from top to bottom (5.8–6.2 m thick) vermicular ferricrete, murram zone, ferruginised pisolitic sandstone band, kaolin rich whitish clay and thin ferruginized sandstone. The ferricrete of 3.7–4.4 m thickness is vermicular and cavernous in nature with interconnected paths of fluid passage. The crust is constituted of highly limonite and goethite cemented coarse to fine quartz and feldspar grains, iron nodules, and concretions. The zone of *murram* is composed of highly lateritized sediments, including sub-angular pebbles and quartz grains which are appeared as stone line to demarcate the separation of ferricrete zone from the saprolite zone. The appearance of whitish kaolin rich clay signifies high degree of lateritization and weathering of sandstone.

4.4   Laterite over Rajmahal Basalt Traps

It is a worth mentioning fact that the typical primary or in situ type laterite profiles with little bauxite enrichment are well developed over the basalts, signifying increasing depth (10–15 m) of basal weathering and forming oldest laterites in the western part of Bengal Basin. Such sections of laterites are observed in details at Sultanpur (24° 22′ 33″ N, 87° 47′ 31″ E), Baramasia (24° 10′ 54″ N, 87° 43′ 05″ E), and Madhupur (24° 17′ 53″ N, 87° 42′ 43″ E) along Rampurhat–Dumka road to understand the genesis of laterite on the Rajmahal Basalts (Early–Late Cretaceous).

4.4.1   Sultanpur Profile

A well-developed and well-preserved laterite profile of about 11 m thick will all its attributes of primary laterites is exposed at Sultanpur (24° 22′ 33″ N, 87° 47′ 31″ E) near Nalhati, Birbhum district (Fig. 4.6). Pisolitic hardcrust with residual ferruginous latosol varies in thickness from 2.55 to 2.75 m. The broken pisolites show core to rim colour banding of limonitic to goethite composition (i.e., gritty layer). About 2.75–4.0 m depth, we have found ferricrete pisolite zone which is characterized by relict columnar structure of basalts. It corresponds to a progressive accumulation of iron and as a consequence, to a progressive development of hematitic iron nodules. The bleached zone is reduced in size, so that the yellow–white coloured domain decrease in size while the purple–red indurated domain enlarges and develops.

A goethite cortex (concentric yellow brown) develops at the periphery of purple–red hematitic nodules. Below the hardcrust, the thick mottle clay horizon with relict columnar structure (4.00–6.75 m) and laminated white kaolinite clay horizon with yellow ochre with small channels of Fe–Al oxides (6.75–9.00 m) are developed.

Sultanpur Profile

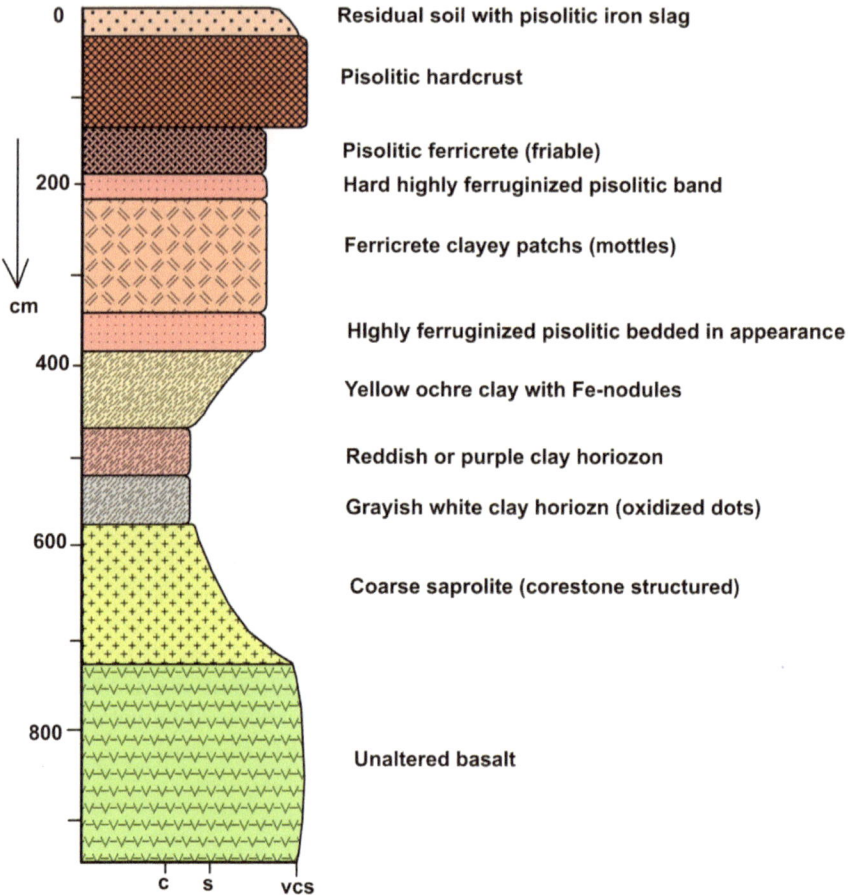

Fig. 4.6 Well-developed sub-surface lithosection of primary laterite on the weathered Rajmahal Basalt at Sultanpur, Birbhum

Fe-mottles, mostly of a brown red colour, are diffuse glaebules and result in a concentration of iron which precipitates mainly as goethite and as haematite together with kaolinite matrix. The dominant minerals are secondary kaolinite [$Al_2Si_2O_5(OH)_4$)] and ferruginous hydroxides in amorphous phase. This is followed by saprolite zone of weathered Rajmahal basalts having liesegang banding and weathering rinds (Fig. 4.4). These trap basalts are spheroidally weathered at base. It is constituted of plagioclase, pyroxene, opaque, and glass with intergranular to intersertal texture. An analogous profile is found at an altitude of 227 m near Ghurnee Pahar, Birbhum (24° 15′ 43″ N, 87° 39′ 11″ E) (Table 4.1) and Pinargaria, Shikaripara, Jharkhand (24° 12′ 13″ N, 87° 40′ 13″ E).

Table 4.1 Account of in situ laterite log profile at Ghurnee Pahar, Birbhum District

Profile depth (m)	Lithological description of horizon
0–0.15	Dismantled gritty layer and slope wash with thin latosol
0.15–2.0	Vermicular hardcrust with tubes and channels, domination of haematite and goethite
2.0–2.15	Highly ferruginized pisolitic haematite band, reddish brown smooth colour
2.15–4.00	Mottle zone with upward moving channels of Fe–Al matrix in bleached purple red kaolinite medium, lithorelictual mottles, columnar structure
4.00–6.15	Purple to greyish yellow kaolinite clay, pallid zone with few channels of ferruginous to aluminous oxides, lithomarge
6.15–6.85	Saprolite, spheriodally weathered basalt, abundance of weathering rinds and liesegang banding, corestone like appearance in kaolinite clay
6.85–9.25	Unaltered basalts (quartz–tholeiites), columnar structure

4.4.2 Baramasia Profile

A section of about 7 m depth from fine saprolite to in situ and ex situ laterite has been studied (Fig. 4.7). A dismantle gritty and slope wash ferruginous sediment layer (globular and elliptical iron pisolites) of 35 cm thick overlies on the 1.1 m thick reworked and re-cemented ex situ laterite which is developed as re-lateritization of the deposited Fe nodules with many petrified wood fossils. This unit is followed downward by very hard and compact iron rich pisolitic ferricrete (1 m thick) with mottled zone at base. The kaolinite cavernous clay layer of 2.3 m thick is yellow ochre clay with few pisolitic nodules and the lowermost 1.5 m thick clay is characteristically greyish–white in colour with oxidized dots of opaque and mafic (cavernous), altering of basalts into fine saprolite. The most interesting feature is that 25 cm thick and hard ferruginized pisolitic band is observed as sandwich between two clay layers.

4.4.3 Madhupur Profile

In a *murram* quarry of Madhupur near Rampurhat, three distinct domains (i.e. lithomarge, mottle zone and duricrust) of laterite profile (Fig. 4.8) are clearly observed, corresponding to the ideal profile of Tardy (1992), Ollier (1995), and Ollier and Sheth (2008).

(a) *Saprolite Domain, or Lithomarge* (>7.2 m depth from land surface)—The saprolite-alteration domain (i.e., strongly weathered basalt) are normally located below the groundwater table (i.e., saturated zone), that is in permanently wet condition, having a depth of more than 11.5 m. In this profile, fine saprolite or lithomarge is observed in between 6.5 and 8.3 m depth and the structures of the parent rock and the original volumes are almost preserved. The dominant min-

Fig. 4.7 Successive development of secondary and primary laterites over kaolinite saprolite at Baramasia of Birbhum district

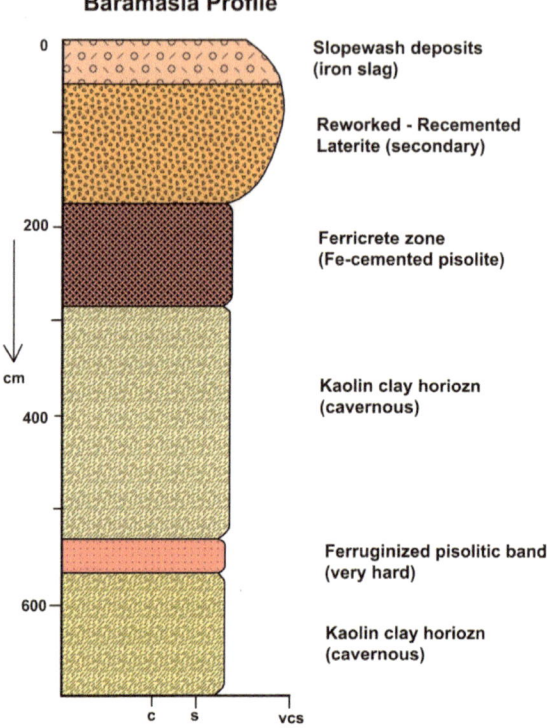

Baramasia Profile

Slopewash deposits (iron slag)

Reworked - Recemented Laterite (secondary)

Ferricrete zone (Fe-cemented pisolite)

Kaolin clay horiozn (cavernous)

Ferruginized pisolitic band (very hard)

Kaolin clay horiozn (cavernous)

erals are secondary kaolinite [Al$_2$Si$_2$O$_5$(OH)$_4$)] and ferruginous hydroxides in amorphous phase. An excessively leached lithomarge (Tardy et al. 1991) corresponds to pallid zone is found here, underlining the mottled zone and separating by red bands of laminated lithomarge and ferruginous grit (Ghosh and Guchhait 2019).

(b) *Glaebular Domain*—According to Ollier (1991), Tardy et al. (1991) and Trady (1992) under contrasted tropical climates fine saprolite, lithomarge and iron are naturally redistributed and concentrated in distinct positions to characterize a glaebular zone, in which duricrust or ferricrete may develop (Fig. 4.3a).

  b1. *Mottle Zone* (7.2–4.1 m depth from land surface)—Fe-mottles, mostly of a brown red colour, are diffuse glaebules and result in a concentration of iron which precipitates mainly as goethite and as haematite together with kaolinite matrix. Due to intensive leaching of kaolinite, macrovoids (tubules and alveoles) are formed. Importantly, lithorelictual Fe-mottles (Tardy 1992; Ghosh and Guchhait 2019) are accumulated, reflecting the palaeostructure of weathered Rajmahal basalt (Fig. 4.9).

  b2. *Ferricrete Zone* (4.1–0.8 m depth from land surface)—It corresponds to a progressive accumulation of iron and as a consequence, to a progressive development of hematitic iron nodules. The bleached zone is

(a) **(b)**

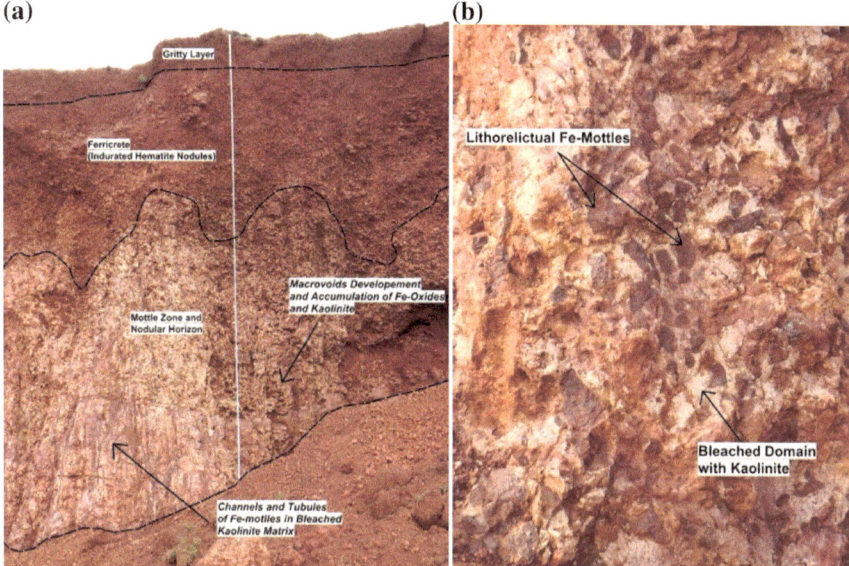

Fig. 4.8 **a** Gritty dismantled layer at top, massive ferricrete at middle and mottle zone with kaolinite matrix of primary laterite profile (6.77 m depth) at Madhupur, Birbhum and **b** a close view of lithorelictual iron mottles in pale-yellowish kaolinite in that profile, reflecting the palaeostructure of weathered Rajmahal basalt (Ghosh and Guchhait 2015)

reduced in size, so that the yellow–white coloured domain decrease in size while the purple–red indurated domain enlarges and develops. A goethite cortex (concentric yellow brown) develops at the periphery of purple–red hematitic nodules. The colours, ranging from yellow to brown, orange–brown and brownish black, signify the presence of limonite which consist of poorly crystalline goethite or lepidocrocite (γ-FeO(OH)) and adsorbed water, i.e., FeO·OH·nH$_2$O (Ghosh and Guchhait 2019).

(c) *Dismantled Gritty Layer* (0.8–0 m depth from land surface)—A gritty horizon is developed at the top laterite profile with development of latosol. This horizon is made of the products of the dismantling of the pseudo-conglomeratic (*Gmg* facies) of the pisolitic-underlined ferricrete. A surfacial sandy layer, made of corroded quartz, is liberated by the dissolution of the ferricrete and a pebbly layer develops at the expense of the pisolitic iron crust, came from the early hematitic nodules (Ghosh and Guchhait 2015).

Fig. 4.9 Development of in situ laterite profile over Rajmahal basalts at Madhupur of Birbhum district

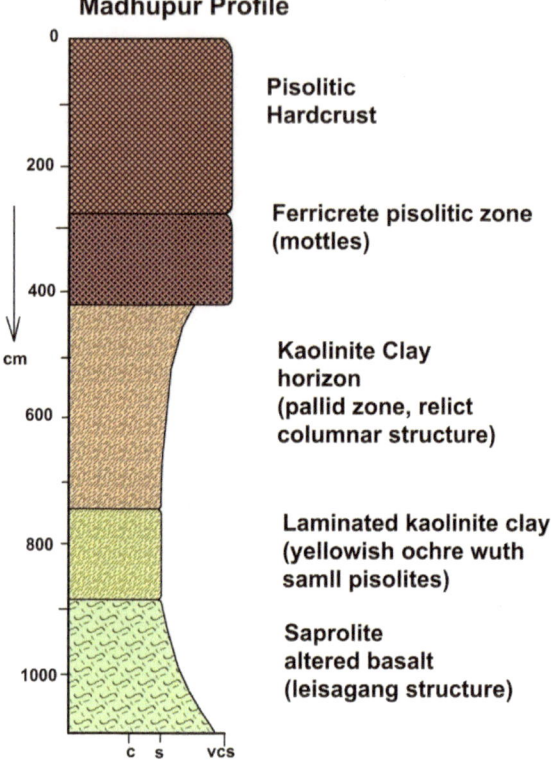

4.5 Justification on In Situ Origin of Laterites

The in situ or primary lateritization processes on the basaltic bedrocks, at Sultanpur (24° 22′ 33″ N, 87° 47′ 31″ E), Baramasia (24° 10′ 54″ N, 87° 43′ 05″ E), Madhupur (24° 17′ 53″ N, 87° 42′ 43″ E), Nalhati (24° 17′ 47″ N, 82° 49′ 28″ E), Pinargaria (24° 12′ 13″ N, 87° 40′ 13″ E), Ichhanagar (24° 22′ 33″ N, 87° 47′ 26″ E), Chaukisal (24° 19′ 59″ N, 87° 41′ 31″ E), Mathurapahari (24° 03′ 03″ N, 87° 36′ 39″ E), and many other lithosections along Rampurhat–Dumka Highway (NH, National highway 114A), are characterized by well-developed laterite profiles (Fig. 4.4), starting from (1) saprolite of weathered basalts, (2) lithomargic clay or pallid zone, (3) mottled zone with litho-relicts of weathered basalts, and (4) ferricrete. The primary laterites on the Chotanagpur gneiss and dolerite dykes are characteristically very thin and continued of poorly developed saprolite zone ferricrete hard crust (Ghosh and Guchhait 2019).

Laterites on the Gondwana sandstones, at Bastabpur (23° 48′ 50″ N, 87° 13′ 29″ E), Hetampur (23° 47′ 03″ N, 87° 23′ 37″ E), Deucha (23° 24′ 06″ N, 87° 14′ 55″ E), Pansiuri (23° 46′ 39″ N, 87° 16′ 47″ E), Dubrajpur (23° 47′ 12″ N and 87° 25′ 19″ E), Ghutgaria (23° 25′ 09″ N, 87° 15′ 11″ E), Barjora (23° 25′ 41″ N, 87° 16′ 32″ E), and Saharjora (23° 24′ 36″ N, 87° 14′ 32″ E), often show ferruginous

(a) **(b)**

Fig. 4.10 **a** The laterite profile of Baramasia showing Fe–Al mottle zone with presence litho-relict structure of weathered basalt and **b** development of hard ferruginized pisolitic band at Sultanpur profile (*Note* Length of scale is 30 cm)

saprolite, ferruginous arenite and thick unit of pisolitic hard crust. A model of in situ development of laterite profile (development of alteration saprolite zone, glaebular mottle-ferricrete zone and the upper soft zone of Fe nodules) is applied to understand the complete formation of laterite layers (Ghosh and Guchhait 2019). Under tropical wet-dry climate of Palaeogene to Neogene, the Rajmahal basalt was weathered to form lithomarge or fine saprolite (i.e., kaolinite) due to intense leaching of silica and chemical alteration of primary mineralsand concurrently the secondary iron and aluminium oxides were cemented towards surface to form ferricrete. The relict structures of weathered basalt are found in the mottle zone, having elongated tubes and channels of haematite, Al-goethite and kaolinite (Figs. 4.10 and 4.11).

The laterites developed over Neogene gravel sediments (mainly in the districts of Bankura and West Medinipur) also indicate in situ lateritization of lower conglomerate, i.e., pebble horizon–siltstone–sandstone units, not the upper older or newer alluvium. The indication of intensive tropical weathering and ferruginous transformation is evidenced from the spheroidal weathering, weathering rinds and liesegang structures of basaltic saprolite (prominent in the lithosections of Nalhati, Ichhanagar, Chaukisal, and Mathurapahari) which are characterized by reddish brown to yellow coloured fine grained clayey core and iron rich chocolate brown coloured rims (Ghosh and Guchhait 2019).

Occurrences of white kaolinite clay horizon (i.e., oxidized dots of opaque and mafic within pallid zone) and reddish or purple colour to yellow ochre clay (with

Fig. 4.11 a Lithosections of dismantled ferruginous layer, vermicular ferricrete and Fe–Al litho-relict mottles at Boro Pahari (24° 11′ 49″ N, 87° 42′ 39″ E), Birbhum, **b** development of secondary lateritic hard crust upon massive primary laterite profile at Bhatina (24° 10′ 02″ N, 87° 42′ 25″ E), Birbhum, **c** weathering rinds and core stones (saprolite) of Rajmahal basalts in weathered medium at Nalhati, Birbhum, and **d** channels of Fe-mottles (litho-relicts of weathered basalts) in kaolinite matrix at Bhatina, Birbhum (*Note* Length of scale is 30 cm) (Ghosh and Guchhait 2019)

ferricrete nodules and pisolites) is the direct product of lateritization process below the profile. At Nalhati and Ichhanagar sections, the development of very hard and compact iron rich pisolitic band (constitutes of feldspar, opaque and pisolites of iron concretions) signifies crudely weathered topmost surface of Rajmahal basalt, having litho-relicts of basaltic texture. The in situ ferruginous hard crust is generally thick vermicular type with well-developed rounded concretions of haematite and relict fluid passage paths (Fig. 4.10).

The zone of red pisolitic ferricrete with channels of mottles (1.5–1.75 m thick) at Sultanpur lithosection represents a true in situ ferricrete of *Rarh Bengal*. The upper dismantled horizon of ferricrete consist of loosen globular to elliptical iron pisolites which show core to rim colour banding of limonitic to goethite composition which reflects cortex development due to exposure of laterite mantle under the tropical wet–dry palaeoclimate. The pisoid structured ferricrete of Bastabpur section (2–2.6 m thick) is very hard and compact in nature and continued of globular pisolites whereas the pisoid structured part is characteristically rich in pisoid and fluid passage paths and channels which acted as pathways for fluid or solution migration (leaching of silica) down the profile.

In general, the in situ hard crust nodules are circular to amoeboid shaped concretions of iron oxides (mainly haematite, limonite, and goethite cortex) and sand balls. Centripetal accumulation of in situ ferruginous materials to sub-nodules and to meta-nodules includes tropical dehydration of haematite, Al-haematite and Al-goethite at the surface in dry period. Observing the primary laterite lithosections of Nalhati and Rampurhat region, we have found few interesting phenomenon about magnetite to haematite nodule formation along the profile. In saprolite zone, due to intense weathering under pressure, magnetite grains are altered to blood red coloured haematite at surfaces. In pallid zone, these grains are mostly altered to haematite pseudomorphs by oxidation around edges and along cracks (as channels or bands). In mottle zone, the biotic grains are completely replaced by goethite in the kaolinite medium and magnetite grains are replaced by haematite mottles, i.e., Fe-litho relicts (Figs. 4.10 and 4.11).

References

Ghosh S, Guchhait S (2015) Chraterization and evolution of primary and secondary laterites in northwestern Bengal Basin, West Bengal, India. J Palaeogeogr 4(2):203–230

Ghosh S, Guchhait SK (2019) Modes of formation, Palaeogene to Early Quaternary Palaeogenesis and geochronology of laterites in Rajmahal Basalt Traps and Rarh Bengal of Lower Ganga Basin. In: Das BC, Ghosh S, Islam A (eds) Quaternary geomorphology in India. Springer, Singapore, pp 25–60

Ollier CD (1991) Laterite profiles, ferricrete and landscape evolution. Zeitschriftfür Geomorphologie 35(2):165–173

Ollier CD (1995) New concepts of laterite formation. Memoirs Geological Society of India, No 32, pp 309–323

Ollier CD, Sheth HC (2008) The high Deccan duricrusts of Indian and their significance for the 'laterite' issue. J Earth Syst Sci 117(5):537–551

Tardy Y (1992) Diversity and terminology of laterite profile. In: Martini IP, Chesworth W (eds) Weathering, soils and paleosols. Elsevier, Amsterdam, pp 379–405

Tardy Y, Kobilsex B, Paquet H (1991) Mineralogical composition of geographical distribution of African and Brazilian peri-Atlantic laterites: the influence of continental drift and tropical paleoclimates during the past 150 million years and implications for India and Australia. J Afr Earth Sci 12(1–2):283–295

Chapter 5
Characterization of Secondary Laterite Profiles

Abstract This section minutely analysed the profiles of low-level or ex situ laterites which separate the lithological formations of Archaean, Gondwana, and Tertiary gravels from the Sijua and Chuchura Formations (Quaternary Alluvium) deposited over the shelf zone of Bengal Basin between Pliocene and Pleistocene. The secondary laterites are specifically found as the dissected interfluves in Rampurhat, Illambazar, Bolpur, Kanksa, Ausgram, Bishnupur, Garhbeta, and Kharagpur through the variations of horizons are still observed. The presence of a subsurface layer of kaolinitic clay reddened from above by ferro-colloids and of rounded pebbles of different rocks, gravelly appearance of crust and the general absence of conventional horizons of laterite signify a detrital origin (ex situ) in *Rarh* Plain.

Keywords Hardcrust · Murram · Ferricrete · X-bedding · Conglomerate

5.1 Occurrences of Secondary Laterites

The badland topography (i.e. *Khoai* landscape) of Bolpur (Kopai–Ajay interfluve) has developed over *murram* (i.e. friable ferruginous concretions) composed of loosely bounded iron concretions, gravels with a rather thin and hardened surface layer and a brunt blackish red colour. In the forest tract of Durgapur, Kanksa, Galsi, Augram I and II blocks of Bardhaman district (Ajay–Damodar interfluve), the harderust of gravelly laterite is partly eroded by rills and gullies. Here, two distinct layers of pisolitic nodules are identified—(1) loosely and less compacted gravelly iron concreted zone (70–80 cm thick) and (2) compacted and hard layer composed of regularly spaced nodules cemented by red clayey matrix being ox-blood in colour. The lateritic exposure in the Silabati–Dwarkeswar interfluve (at Bishnupur) is appeared as 'laterite mesa' with 10-m relative relief. The remnants of hardcrust have a thickness of 3 m, and it is overlain on mottle zone and whitish or light yellow kaolinitic pallid zone. Spectacular gully development at Gangani near Garhbeta, West Medinipur district, has given rise to a micro-level badland development over the multi-level laterite profile near the right bank of Siliai River. In the following section, the detailed morphological studies of laterite profiles are discussed at Radha Damodarpur

(23° 04′ 15″ N, 87° 22′ 23″ E), Gangani (22° 51′ 21″ N, 87° 20′ 45″ E), Nanda Bhanga (23° 10′ 31″ N, 87° 15′ 51″ E), Durgadaspur (22° 32′ 00″ N, 87° 19′ 10″ E), Jamsol (22° 25′ 54″ N, 87° 14′ 22″ E), Kharagpur (22° 17′ 37″ N, 87° 15′ 10″ E), Illambazar (23° 36′ 53″ N, 87° 32′ 00″ E), Surul (23° 40′ 18″ N, 87° 38′ 52″ E) and Kharia (23° 59′ 11″ N, 87° 35′ 31″ E).

5.2 Laterite over Late Tertiary Sediments

5.2.1 Radha Damodarpur Profile

In the exposed profile of 6 m thick along a stream and associated area, the ex situ laterite is found over the conglomerate and sandstone unit (Fig. 5.1). The laterite is overlain successively by older alluvium and newer alluvium sediments. The lowermost bedded sandstone unit of 1.2 m thick is more indurated and lithified with a clayey alteration. It is coarse-grained and also granular (quartz, feldspars, mica, and opaque), yellowish white in colour and characterized by kaolin-rich clayey formation. This sandstone is sub-arkosic with feldspar content which is 10–15% and the quartz content varies from 80 to 90%. The sandstone unit represents the Pliocene age. This unit is un-conformably overlain by a ferruginized conglomerate unit (1.3 m thick) which is oligomictic in character with pebbles of dominant quartz and little feldspar. It is bedded in nature and at places show crude large-scale X-bedding. It is almost a clast-supported conglomerate with ferruginized coarse sand-sized matrix of predominantly quartz and Fe nodules. The clasts are varying size from 2 to 13 cm. The matrix is highly cemented by ferruginous oxides (mainly limonite and goethite) which make it very indurated. The clast composition and character indicate its provenance to be vein quartz and pegmatite bodies, and it has suffered a long transportation. Thus, this basal conglomerate unit is a modified and re-cemented duricrust which is secondary in nature and it lies un-conformably on Tertiary sandstone.

The upper part of this conglomerate unit is represented by *murram* with quartz pebbles (1–3 cm in size) and Fe–Al nodules in a ferruginized sand-sized matrix. The lateritized sands are amoeboid shape and at places characterized by vermicules. In this profile, the ferruginous oligomictic and X-bedding conglomerate unit and *murram* form the secondary laterite unit. This laterite unit is overlain by caliche impregnated grey–ash coloured loamy sediment which represents older alluvium of Late Pleistocene age and the yellowish–white coloured sandy newer alluvium overlies the older alluvium.

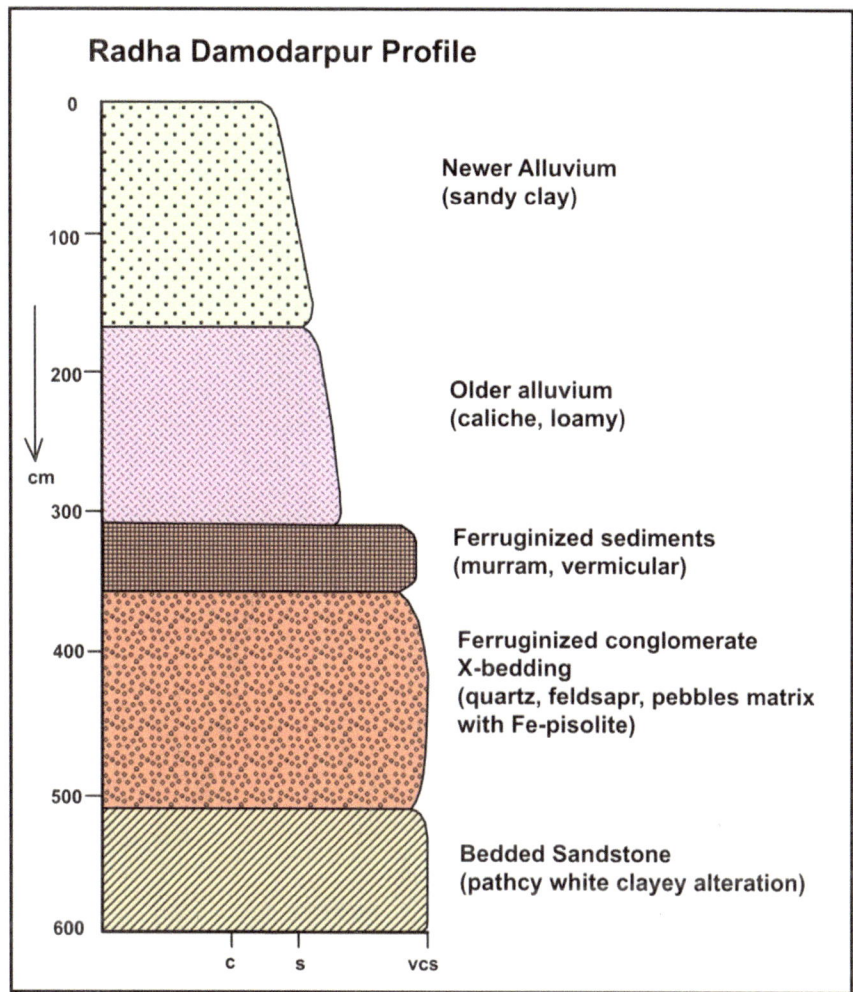

Fig. 5.1 Development of ferruginized sediments in between older alluvium and conglomerate bed at Radha Damodarpur of Bankura

5.2.2 Gangani Profile

About 12–14-m-thick package of alternate sandstone–siltstone sequence (representing fluvial cycles of sedimentation) with overlying ferruginized sediments of fining-upward matrix exposes at the right bank of Silai River of Gongoni, Paschim Medinipur (Figs. 5.2 and 5.3). The exposed vertical section of about 14 m may be divided into two geological units. The lower unit is mainly constituted of alternate sandstone–siltstone lithologies whose contacts at several levels are marked by thin hard ferruginized sandstone layers of 4–8 cm thick. The siltstone unit is very finely

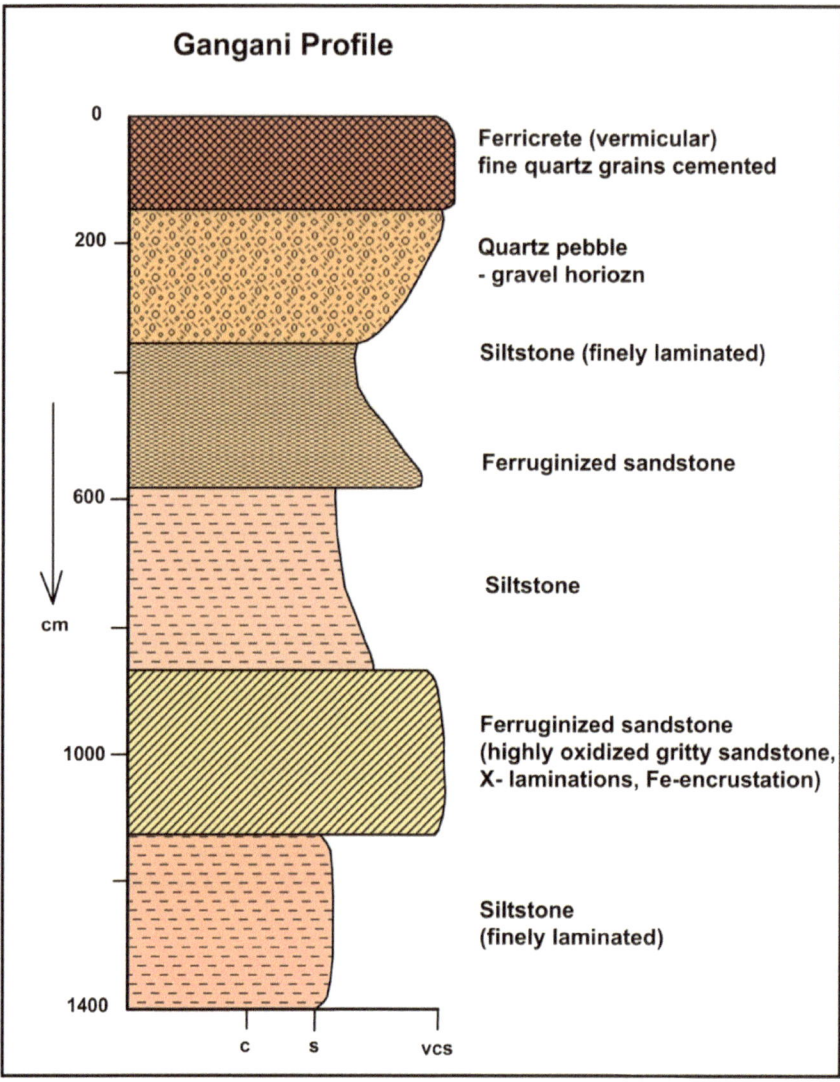

Fig. 5.2 Development of multi-level ferruginized hardcrusts on Tertiary sediments at Gangani, West Medinipur

laminated yellowish white and purple white in colour and at places show whitish patchy clayey alteration.

The lowermost sandstone unit occurring at depth in between 8.5 and 11.7 m is reddish in colour, highly oxidized, and characterized by high degree of iron encrustation features. It is coarse-grained gritty in character and at places few quartz pebbles, with primary sedimentary structures including cross-bedding, crude-graded bedding

Fig. 5.3 Development of multi-level laterite profiles over sandstone–clay bed sequence at Gangani, Wet Medinipur

(cross-lamination and planer bedding), erosional facies, etc., are displayed typically. Another feature includes hollow-pipe-like structure of 2.5–4 cm diameter which is made up of coarse gritty sandstone and developed in different angles even some of which are bifurcating in nature. It is recognized that this ferruginized unit can be marked as second profile of ex situ laterite, i.e. highly oxidized sandstone.

The highly ferruginized thin layers (4–8 cm thick) occurring in between sandstone and siltstone units are constituted of chocolate-coloured, Fe-oxide-cemented quartz grains—a highly ferruginized quartz arenite. This sandstone–siltstone units together may be clubbed as Late Tertiary fluvial sediments (3.7–14 m depth). Along the exposed vertical section, the top 2–3.7 m of lateritized sediment belongs to Early Quaternary age. The top laterite is characterized by ex situ ferricrete, followed by *murram* zone and lowermost quartz pebble horizon (Fig. 5.3). The zone of quartz pebbles (sub-angular to sub-rounded, 2–4 cm in size) un-conformably overlies the Tertiary siltstone unit. The uppermost ferricrete unit is vermicular and blocky in nature with numerous fluid passage paths. The zone is constituted mainly of finer quartz grains and cemented by goethite and limonite. The upper ferruginous layer of this section is the re-lateritization of fluvial derived sediments with upward coarsening sequence. It can be said that there are multi-level lateritized sediments (signifying different episodes of lateritization processes), developing a typical example of secondary laterite profile of Late Tertiary–Early Pleistocene.

5.2.3 Nanda Bhanga Profile

In this area, a cumulative 6-m-thick profile section of sediments has been analysed to exemplify the occurrence of secondary laterites (Fig. 5.4). The profile is classified into two major distinctive litho-units—(i) the lower unit (1.4–7 m depth) is composed of cross-bedded and cross-laminated coarse to gritty sandstone, pebble horizon and siltstone, and (ii) the upper unit (0–1.4 m depth) is constituted of coarser and much harder ferruginized sediments. A horizon of 40–50-cm-thick quartz pebbles demarcates the boundary between lower and upper units, looked as stone-line (though fluvially deposited).

The lower unit is characterized by alternate coarse gritty sandstone and pebble horizon, preserving cross-bedded and cross-laminated structure in sandstone. The pebble horizon (sun-angular to sub-rounded shape) preserves crude cross-bedding in a rare instance. These alternate gritty sandstone–pebble horizons with primary structures indicate different regimes of fluvial energy. This zone is overlain by cream-coloured finely laminated siltstone which shows patchy whitish clayey alteration. This fining upward repetitive cycle of fluvial sediments with moderate lithification represents Late Tertiary sediments.

The hardcrust unit of 70–80 cm thickness is represented by relatively loosened iron concretions. The iron concretions are amoeboid shaped and constituted of iron-cemented finer quartz grains. The duricrust is vermicular type, cavernous and very hard and compact, reddish-black in colour and characterized by channels of fluid passage paths which are branching in character. The ferruginized quartz pebble with fining upward sand-sized grains' sequence represents the ex situ lateritization of fluvial sediments. The duricrust preserves Fe cementation of fluvially derived quartz grains with occasional gravels, having Early–Late Quaternary age of formation. Additionally, 25–50 cm latosol (relict fragments of duricrust) has been developed on the top of profile, supporting the vegetation of this area.

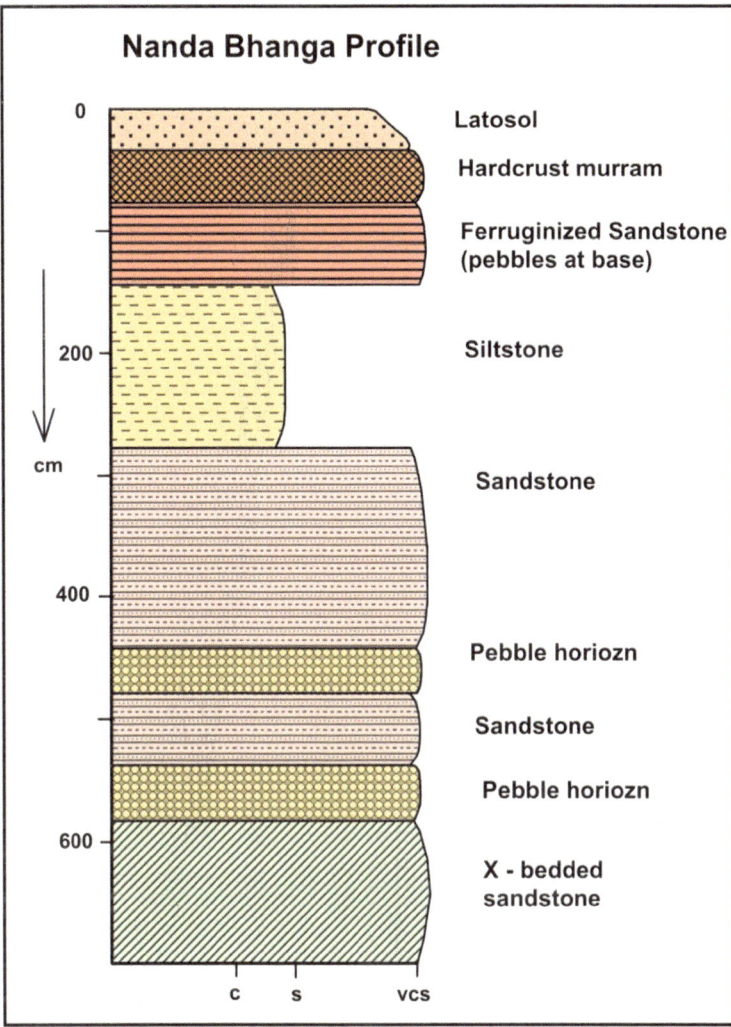

Fig. 5.4 Development of secondary laterite profile over X-bedded sandstone at Nanda Bhanga, West Medinipur

5.3 Laterites over Early Quaternary Sediments

5.3.1 Durgadaspur Profile

A detrital laterite profile of 3–4 m is found in a shallow quarry of Durgadaspur site (Fig. 5.5). The bottom set of profile is represented by coarse sandstone (>1.5 m thick) which shows alumina-rich alteration due to the precipitation of kaolin-rich clay. This sedimentary unit of fluvial origin is successively overlain by a pebble

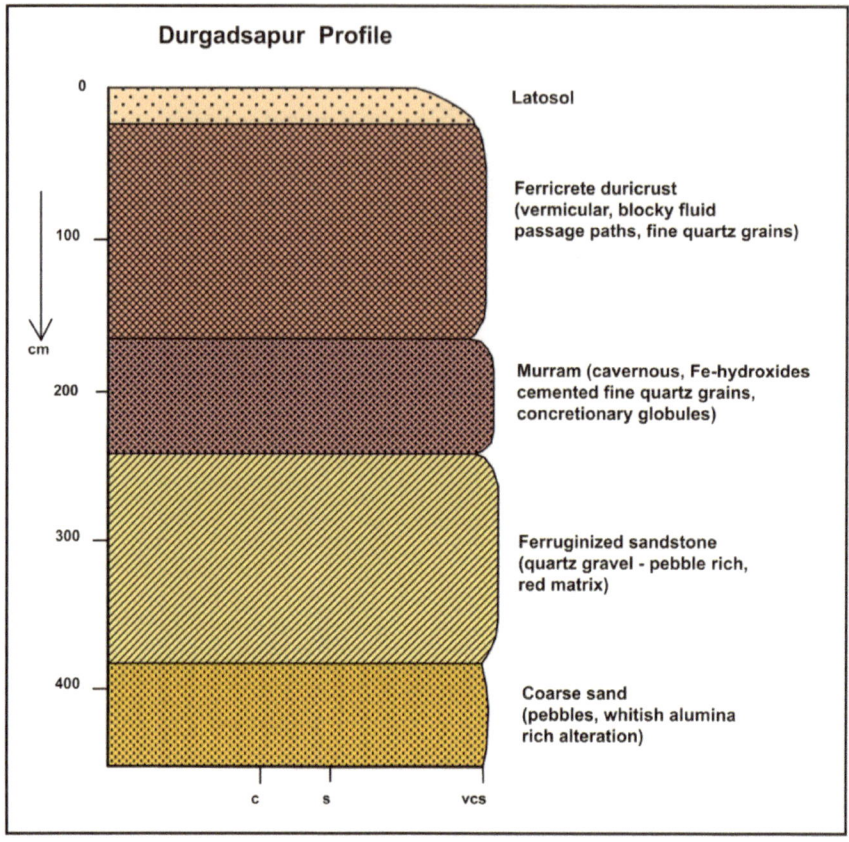

Fig. 5.5 Development of laterite over sandstone at Durgadaspur, Bankura

and gravel horizon, showing upward coarsening of matrix (deposited in high fluvial regime). The quartz-pebble-rich and ferruginized sandstone (1.3 m thick) is cavernous and iron-hydroxide-cemented matrix of fine quartz grains, showing glimpses of Fe concretionary globules.

The layer of *murram* (0.7 m thick) is characterized by strong iron-rich concretionary globules, some irregular-shaped pisolitic globules, cavernous fragments of hardcrust and Fe-hydroxide-cemented gravels. The bottom of *muuram* layer is composed of 1–2-cm-sized sub-angular pebbles. Above this layer, the ferricrete (1.3 m thick) occurs as vermicular type, reddish-brown-coloured and blocky with paths of fluid passage in several directions (showing high degree of leaching of silica). A transitional curvilinear contact is marked at the *murram*-hardcrust boundary. Main constituents of ferricrete are highly oxidized sands, ferruginized quartz grains and fragmented Fe–Al and Fe–Mn nodules. About 30-cm-thick reddish-brown-coloured latosol is developed to support grass-like vegetation. Therefore, it can be said that this laterite profile also categorized as ex situ or secondary type because the profile is formed due to re-lateritization of fluvial derived ferruginous materials which have unconformable contact with Early Quaternary pebble-rich sandstone.

5.3.2 Kharagpur Profile

Widespread lateritic cap rock (found as the resistant layer, Butte type) is exposed over a vast area at the south of Kharagpur town, overlying on the Quaternary sedimentary pebble–gravel–sand deposits. In this site of quarry, about 2.3-m-thick section shows five distinct facies—(i) upward finding sand layer at base, (ii) coarse sand with pebble, (iii) ferruginized cross-bedded repetitive pebble–sand, (iv) ferricrete duricrust, and (v) latosol at top.

The exposed lowermost sand unit is whitish yellow in colour, coarse sand sized with occasional larger clasts and shows some iron staining. This sedimentary layer has not any glimpses of weathering alteration and relation with the upper ferruginous layers. The *murram* (1 m thick) represents here in the form of ferruginized pebble–sand layers which are cross-laminated in several places, with iron encrustations. The ferricrete duricrust (0.65 m thick) is vermicular type, hard, and blocky in nature. It is composed of goethite- and limonite-cemented fine-to-coarse sand-sized clasts which are the fluvial sediments and slope wash sediments. The ferricrete shows branching and non-branching types of fluid passage which is the path of leaching. About 30–35-cm-thick latosol is developed, constituting relict duricrust and quartz grains. The important facts of this laterite section are that the effect of laterization is observed up to 1.9 m beyond which only iron staining on quartz grains are noticed and the ferruginous facies are developed by the re-cementation of deposited ferruginous materials with quartz grains.

5.3.3 Jamsol Profile

An exposed sedimentary section of 6 m thickness with lateritized sediments is examined at the left bank of Kasai River (Jamsol, West Medinipur) (Fig. 5.6). The lithosection represents the lowermost cross-stratified and well-lithified sandstone unit (1.3 m thick), massive coarse sand with hard ferruginized sandstone bed (1 m thick), sandstone with kaolinite (1.2 m thick), hard ferruginized sandstone (0.85 m thick), pebble horizon with sand matrix (0.95 m thick), ferricrete duricrust (0.4 m), and latosol (30 cm).

The lower four units of fluvial deposition, i.e. from cross-bedded sandstone to ferruginized sandstone (1.65–6.0 m depth), belong to Late Tertiary formation. The units had undergone low-to-medium level of ferruginous alternation, forming Fe–Al oxides matrix of coarse sand and gravel. The upper ferruginous unit shows upward fining sequence of fluvial sediments with high degree of lateritization. The upward fining pebble horizon shows weak matrix of Fe nodules and oxidized sands, showing

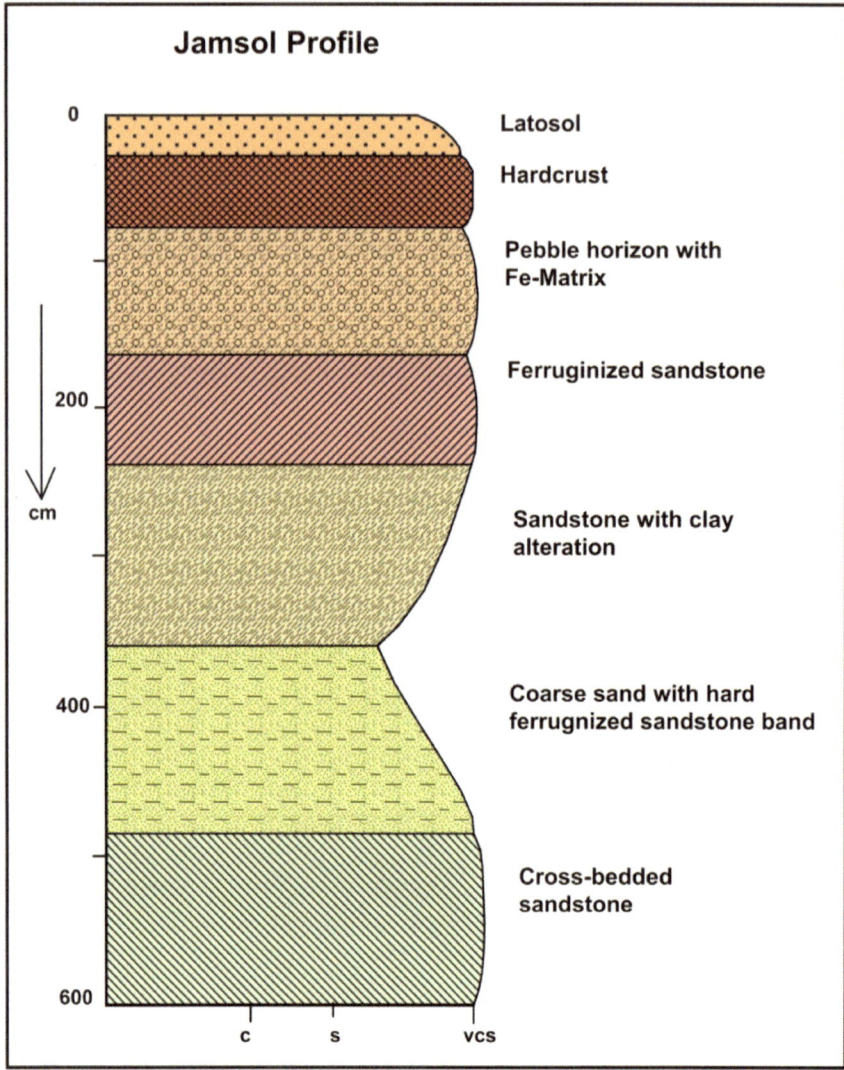

Fig. 5.6 Development of ex situ laterite over sandstone–clay–pebble deposits at Jamsol, West Medinipur

facies of fan-deltaic deposition and re-lateritization. The hardcrust shows vermicular type with fluid passage paths and hard–compact cementation of deposited Fe nodules. The hardcrust appears as the cap rock to resist the underlying sedimentary unit from fluvial erosion. In this profile, two distinct units of ex situ lateritization can be recognized—(i) first-level Tertiary sedimentary unit with ferruginized sandstone from 2.4 to 6.0 m and (ii) second-level Quaternary lateritization unit up to 2.4 m depth.

5.3.4 *Ilambazar Profile*

At the left bank of Ajay River, near Ilambazar of Birbhum, the laterite profile is characterized by 1.35-m-thick reddish-black vermicular hardcrust followed by 60-cm-thick pisolitic ferricrete zone with sand matrix and 75-cm-thick mottle zone (Fig. 5.7a). The hardcrust is vermicular type and re-cementation of channel-fill deposits, composed of highly ferruginized pebble horizon and petrified wood fossils in a ferruginized sand matrix. The iron cement is characteristically limonitic and goethitic. Ferricrete pisolitic zone is less reddish in colour and constituted of iron pisolites along with sands and lowermost mottle zone is a yellowish-white-altered zone

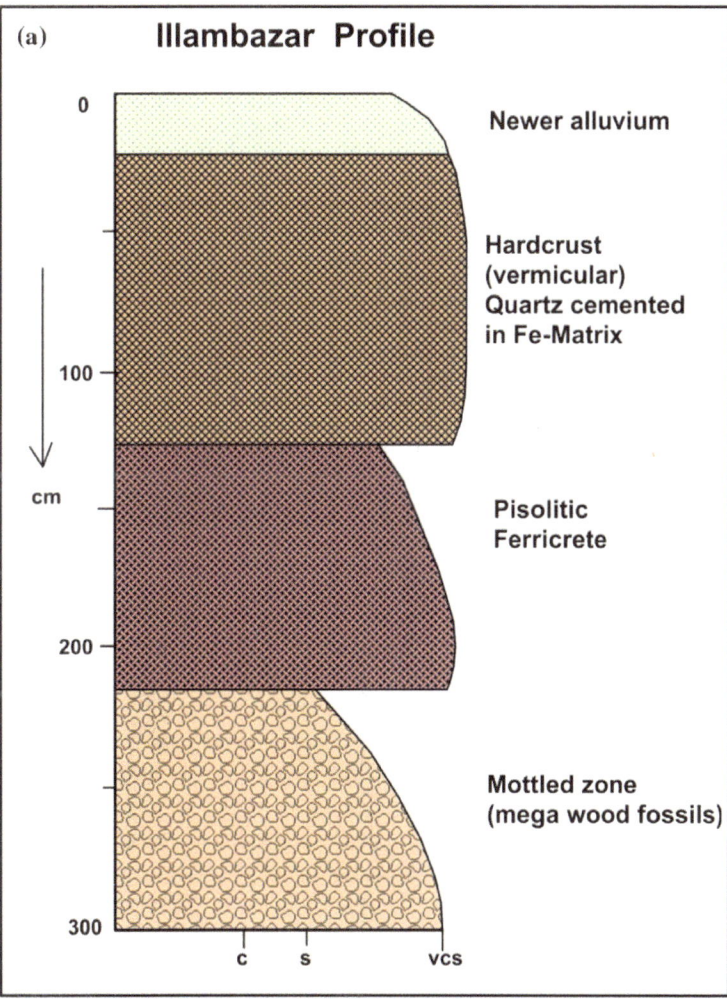

Fig. 5.7 Development of ex situ laterite profiles at **a** Illambazar and **b** Surul of Birbhum district

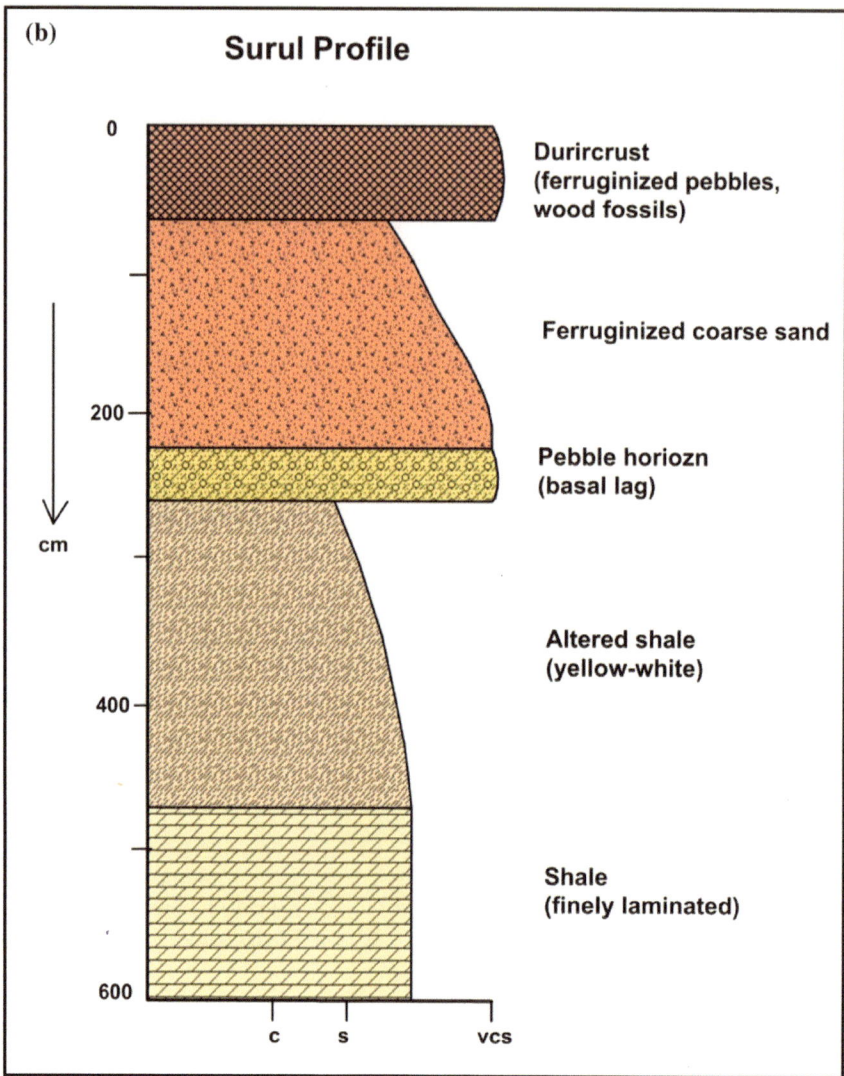

Fig. 5.7 (continued)

of sandstone at places patchy clayey in appearance. The laterite section is covered by newer alluvium (1–1.25 m thick) of Ajay River; but due to bank erosion, it is exposed to the surface. The similar laterite sections are observed at Jambuni and Kheliadanga sites of Birbhum. In this region, the ferruginous duricrust of gravel-containing Fe-matrix is un-conformably developed on the fluvial sedimentary units of gravels and pebbles which preserve large-scale cross-bedding with truncated top and asymptotic bottom. The similar profile is also observed at clay quarry of Surul (Fig. 5.7b).

5.4 Justification on Ex Situ Origin of Laterites

The re-lateritization of transported ferruginous materials (i.e. ex situ laterites) is mainly observed in the eastern part of *Rarh* Plain, especially in other lithosections of Bolpur (23° 40′ 18″ N, 87° 39′ 10″ E), Sriniketan (23° 41′ 31″ N and 87° 40′ 31″ E), Labhpur (23° 48′ 47″ N, 87° 46′ 57″ E), Rampurhat (24° 11′ 44″ N, 87° 44′ 03″ E), Illambazar (23° 36′ 56″ N, 87° 32′ 07″ E), Kanksa (23° 28′ 45″ N, 87° 27′ 45″ E), Panagarh (23° 27′ 10″ N, 87° 31′ 51″ E), Patryasayer (23° 12′ 35″ N, 87° 31′ 17″ E), Bishnupur (23° 05′ 28″ N and 87° 16′ 15″ E), Garhbeta (22° 51′ 34″ N and 87° 20′ 28″ E), Khemsuli (22° 20′ 26″ N, 87° 11′ 04″ E), Rangamati (22° 24′ 42″ N, 87° 17′ 55″ E), etc. In Garhbeta section, alternative kaolinite clay and sandstone–siltstone sequences are overlain by some lenticular channel-fill deposits, consisting of various size fragments of quartz grains, pisoids of in situ laterites and larger or out-sized clasts with petrified woods (Ghosh and Guchhait 2019).

The conglomeratic ferricrete hard crust with ample gravels (without horizon of mottle clay, pallid zone and saprolite) at Bishnupur sand Sriniketan sections reflects the secondary origin of laterites. At the section of Bishnupur Neogene sandstone facies is un-conformably overlain by a ferruginized conglomerate unit which is

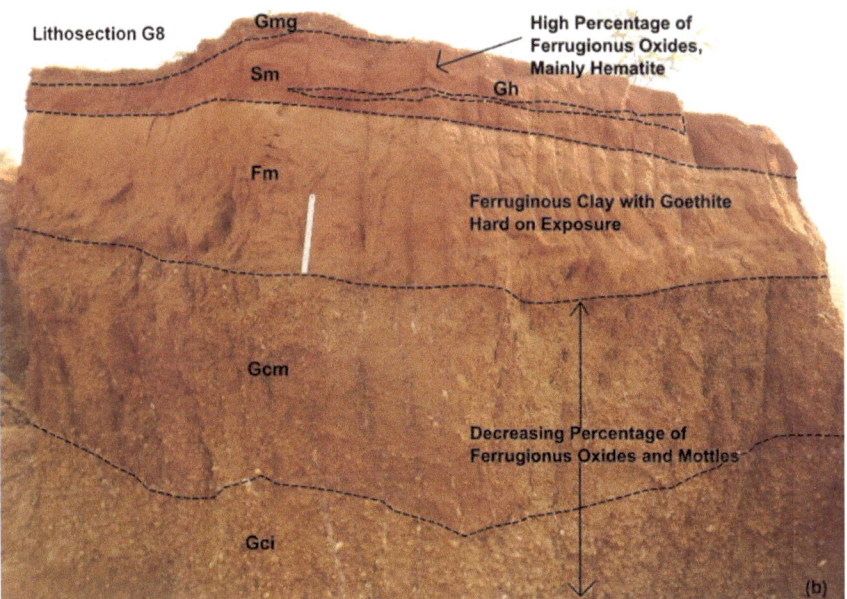

Fig. 5.8 An ex situ laterite profile (2.7 m depth) on the Tertiary gravels at Kamalpur, north of Durgapur, having fluvial to debris flow facies—Gci (i.e. inverse grading clast-supported gravels), Gcm (i.e. Clast-supported massive gravel), Fm (i.e. Overbank or waning flow clay deposits), Sm (i.e. sand massive or faint wavy lamination), Gh (i.e. Clast-supported, crudely bedded gravel), and Gmg (i.e. inverse to normal grading matrix-supported massive gravels) (*Note* length of scale is 30 cm)

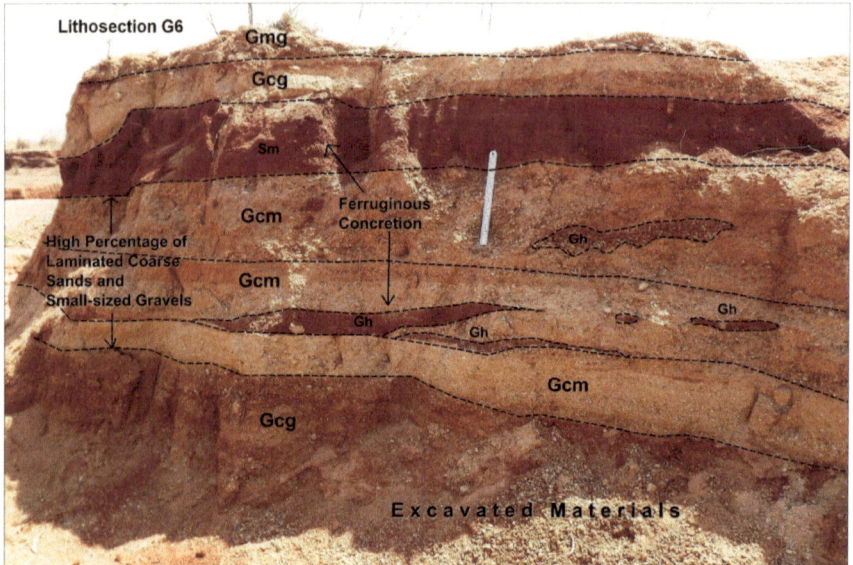

Fig. 5.9 A secondary laterite profile (3 m depth) on the Tertiary gravels, showing alluvial fan to fan-deltaic depositions at Hetodoba north of Durgapur (*Note* length of scale is 30 cm)

oligomictic in character with pebbles of dominantly quartz which is fluvially eroded remnants of primary in situ laterites (Ghosh and Guchhait 2019). The crude large-scale X-bedding sandstone unit of Bishnupur is almost a clast-supported conglomerate with ferruginized sand-sized matrix of predominantly quartz and feldspar. These quartz clasts are mostly well rounded to sub-rounded, smooth surface and flat without any striations which reflects fluvial origin. From the samples of ex situ secondary *laterites*, it is observed that the matrix or groundmass is highly cemented by ferruginous oxides (limonite and goethite) which make it very indurated and the clasts show some iron staining on their surfaces (Figs. 5.8 and 5.9), signifying long-term retention in iron oxides' medium. The clast composition and character indicate its provenance to be vein quartz and pegmatite bodies, and it has suffered a long transportation (Ghosh and Guchhait, 2019).

At Gangani, Garhbeta, and Rangamati sections, the occurrences of ferruginized coarse sandstone (X-laminations and Fe-encrustation), quartz pebble horizon and topmost vermicular hard crust with quartz grains and clastic components again confirmed the re-lateritization episode of deposited materials up to Late Pleistocene (Fig. 5.10). In the profile of Sriniketan section, the degree of ferruginization increases upward and leads to ferruginized coarse sand whose lowermost part is almost free of ferrugination, signifying secondary weak lateritization with less leaching of silica. The variable pisoid shapes of secondary laterites are constituted by two elements—(1) a nucleus of variable nature surrounded by a cortex composed of concentric limonite and goethite and (2) pisoid of mono-nucleus and bi-nuclei composed of sand and clay.

Fig. 5.10 Miocene to Eocene dated dicotyledonous fossil woods found **a** at a laterite quarry of Maluti, Shikaripara (Jharkhand) and **b** at a gully bed of Bhatina, Birbhum, **c** gravel lithofacies of ex situ ferruginous hard crust at Hetodoba, Bardhaman, and **d** progressive badland development on the terrain of secondary laterites and gravel litho-units at Kanksa, Bardhaman

The large-scale cross-bedding of ferruginous sandy unit with truncated top and asymptotic bottom indicate its sedimentary fluvial origin of the sediments, mostly in the palaeofan-deltaic sequences (*Gmg* facies) (Ghosh and Guchhait 2019). Lateritization climate of Paleocene, Eocene and post-Mio-Pliocene favoured good growth of broad leaf tropical and subtropical flora. The occurrences of large-scale petrified dicotyledonous fossil woods (probably Miocene to Eocene age) with ferricrete nodules and channel lag deposits in the ex situ profiles bear the evidence of secondary lateritization up to Late Pleistocene (as OSL data suggested) (Fig. 5.11). Again, this paleontological data suggests the primitive development of ferruginous crust (in situ laterite) in that period of Neogene.

Fig. 5.11 a Development of pisolitic hard crust with stone-lines of gravels and pebbles on kaolinite at Baramasia, Birbhum, **b** nodular secondary ferricrete on left bank of Ajay River at Illambazar, Birbhum, **c** lithofacies of gravels and ferruginous hard crust at Hetodoba, Bardhaman, and **d** formation of secondary lateritic hard crust on kaolinite at Bhatina, Bibhum

Reference

Ghosh S, Guchhait SK (2019) Modes of formation, Palaeogene to Early Quaternary Palaeogenesis and geochronology of laterites in Rajmahal Basalt Traps and Rarh Bengal of Lower Ganga Basin. In: Das BC, Ghosh S, Islam A (eds) Quaternary geomorphology in India. Springer, Singapore, pp 25–60

Chapter 6
Geochemical Properties and Lateritization Processes

Abstract The laterite described by Buchanan is only one member of laterite families whose members have different properties but are similar genesis. In the *Rarh* Plain of GBM delta, the processes of primary and secondary laterite formation are slightly different in the basis of magnitude of involving factors (i.e. type of weathered materials, source of ferralitic materials, wet–dry type of climate, fluctuation of groundwater table, topographic positions, stability of favourable environment, etc.). In the tropical geoclimatic settings, the processes of lateritization (transfers of Fe), latosolization (residual accumulation of Fe), desilication (loss of silica from the profile), and rubification (reddening the regolith and soil horizons with iron oxides) are simultaneously operated to develop distinct horizons of laterite. In this section, various chemical properties of laterite samples, lateritization processes, and applicable theory of lateritization are discussed to get ideas about the genesis of ferruginous layers in the weathering profiles.

Keywords Lateritization · Molar ratio · Triangular diagram · Chemical diffusion · Residuum theory

6.1 Geochemical Analysis

Five samples belonging to pisolitic hard crust (PL 1), ferricrete (PL 2), mottle clay and kaolinite (PL 3) and saprolite (PL 4 and 5) zones are analysed to characterize the chemical variations along the Nalhati lithosection with depth. The result of geochemical analysis (Table 6.1) depicts that sample no. PL 1 belongs to pisolitic crust of Rajmahal Trap laterites and represented by high Al_2O_3 (36.7%), Fe_2O_3 (26.2%) and TiO_2 (4.77%) and very low SiO_2 (9.7%), low MgO (0.14%), CaO (0.17%), Na_2O (0.01%), K_2O (<0.02%) and P_2O_5 (0.11%), respectively (Ghosh and Guchhait 2019). So, it can be said that chemically the most mobile elements such as Si, Na, K, Mg, and P have been removed from the upper part of laterite profile through leaching and solution system and enrichment of Fe, Al, and Ti has been taken place in the top ferricrete zone.

© The Author(s), under exclusive license to Springer Nature Switzerland AG 2020   83
S. Ghosh and S. K. Guchhait, *Laterites of the Bengal Basin*,
SpringerBriefs in Geography, https://doi.org/10.1007/978-3-030-22937-5_6

Table 6.1 Geochemical properties of primary laterite lithosection in Nalhati, Birbhum

Sample no.	PL 1	PL 2	PL 3	PL 4	PL 5	PL 6
Depth of sample from the top of profile (m)	0.65	3.70	5.85	8.25	9.15	9.60
SiO_2 (%)	9.70	28.50	31.46	29.61	31.58	37.67
Al_2O_3 (%)	36.71	26.10	26.42	24.3	24.45	31.15
Fe_2O_3 (%)	26.2	27.86	23.12	27.17	25.33	9.31
MnO (%)	0.29	0.33	0.11	0.12	0.10	0.03
MgO (%)	0.14	0.28	0.06	0.04	0.06	0.24
CaO (%)	0.17	0.07	0.16	0.11	0.18	0.21
Na_2O (%)	<0.01	<0.01	<0.01	<0.01	<0.01	<0.01
K_2O (%)	<0.01	0.02	<0.01	<0.01	0.01	<0.01
TiO_2 (%)	4.77	2.81	3.42	3.69	4.53	4.22
P_2O_5 (%)	0.11	0.11	0.19	0.26	0.09	0.07
Molar ratio	0.14	0.50	0.59	0.54	0.58	0.84

Source Ghosh and Guchhait (2019)

Sample PL 5 represents saprolite or weathered basalt composition as reflected by SiO_2 (31.58%), Al_2O_3 (24.45%), Fe_2O_3 (25.33%), and TiO_2 (4.53%). The low values of other oxides may be the direct result of removal from the system during lateritization. The laterite samples from ferricrete pisolite zone (PL 2), clay horizon (PL 3), laminated clay horizon (PL 4), and saprolite zone (PL 5 and PL 6) indicate low fluctuations of SiO_2 (28.5–31.58%), Al_2O_3 (24.3–26.1%) and Fe_2O_3 (23.12–27.86%) which signifies intensive deep basal chemical weathering of Rajmahal basalt, forming the in situ primary laterite lithosection at Nalhati (Ghosh and Guchhait 2019).

Four samples of Sriniketan lithosection are collected from different horizons of lateritized sediments and siltstone for chemical analysis (Table 6.2). Only, the sample SL 1 shows some enrichment of Fe_2O_3 and lowering of SiO_2 relative to other sediments. High percentage of SiO_2 (60.83–73.39%) reflects low-intensive leaching of silica from the profile. It can be interpreted that the sediments had not undergone any lateritization to develop ideal profile of laterites except ferruginization at the top with coarse sand and gravels. So, this sediment profile with several inputs of earlier laterite represents a reworked secondary laterite lithosection. The ferruginous eroded mantles were fluvially derived from the upper primary laterite zones.

6.2 Lateritization Index

Under hot and sub-humid tropical climate, the prime process of lateritization (i.e. tropical weathering) is the downward leaching of silica and bases (more mobile elements) and upward enrichment of Fe–Al oxides (less mobile elements) (Ghosh and Guchhait 2019). The molar ratio (Birkeland 1984), i.e. index of weathering,

Table 6.2 Geochemical properties of secondary laterite lithosection in Sriniketan, Birbhum

Sample no.	SL 1	SL 2	SL 3	SL 4
Depth of sample from the top of profile (m)	0.35	1.50	3.10	4.45
SiO_2 (%)	60.83	73.39	60.01	58.79
Al_2O_3 (%)	8.88	15.74	20.94	18.44
Fe_2O_3 (%)	20.56	4.12	7.11	7.87
MnO (%)	0.42	0.03	0.03	0.03
MgO (%)	0.82	0.52	1.31	2.32
CaO (%)	0.01	0.20	0.36	0.47
Na_2O (%)	0.05	0	0.04	0.05
K_2O (%)	0.92	2.34	2.67	3.21
TiO_2 (%)	0.41	0.44	0.89	0.82
P_2O_5 (%)	0.09	0.02	0.03	0.04
Molar ratio	2.04	3.61	2.07	2.17

Source Ghosh and Guchhait (2019)

is applied here to depict the degree of leaching. There are marked differences of molar ratio in between the sections of Nalhati and Sriniketan. In Nalhati section, the low value of molar ratio (0.14–0.84) signifies well-developed leaching of silica and segregation of sesquioxides nodules throughout the profile. Molar ratio of ferricrete (sample from 0.65 m depth) is very low (0.14) due to the prevalence of Al_2O_3 (36.71%) and Fe_2O_3 (26.20%), whereas it increases up to 0.58 in kaolinite horizon or lithomarge (sample from 9.15 m depth) due to dominance of SiO_2 (31.58%). Down the profile under acidic environment of tropics, most of smectite and interstratified clay minerals have been transformed to kaolinite, iron oxides and minor amounts of gibbsite (Ghosh and Guchhait 2019).

On the other side, the high value of molar ratio (2.04–2.17) is found in Sriniketan section. The molar ratio of ferruginous crust (sample from 0.35 m depth) is very high (2.04) due to high amount of SiO_2 (60.83%) than Fe_2O_3 (20.56%). Below that crust layer (i.e. ferruginous coarse sand and pebbles at depth of 1.50 m), the molar ratio reaches up to 3.61 due to ample occurrence of SiO_2 (73.39%). In Sriniketan section, the values of molar ratio suggest the relatively very low leaching of silica from the profile and low enrichment of sesquioxides. On the basis of molar ratio, it can be said that the in situ intensive lateritization is observed in the lithosection of Nalhati and ex situ weak lateritization is observed in Sriniketan section (Ghosh and Guchhait 2019).

The triangular diagram (Schellman 1986; Pain and Ollier 1996; Meshram and Randive 2011) shows a high degree of lateritization in the samples primary laterites (Nalhati, Birbhum) and very weak degree of lateritization (preservation of quartz fragments) in the samples of secondary laterites (Sriniketan, Birbhum) (Fig. 6.1). Getting amount of Al_2O_3, CaO, Na_2O, and K_2O from the samples of laterite profiles, we have found that Chemical Index of Alteration (CIA) for primary laterites (Nalhati,

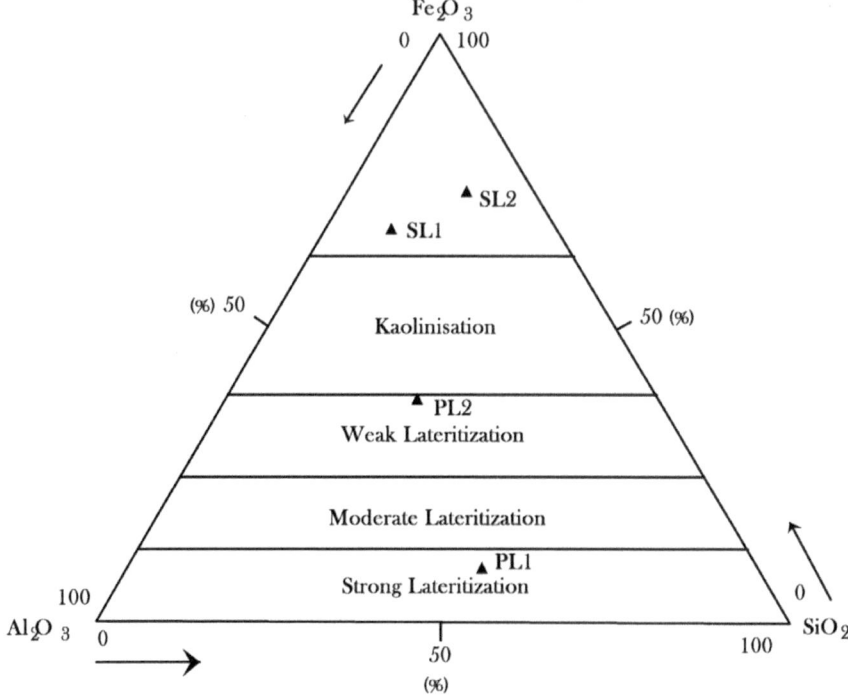

Fig. 6.1 Al_2O_3–Fe_2O_3–SiO_2 triangular diagram showing degree of lateritization in the study area (Ghosh and Guchhait 2019)

Birbhum) is greater than 99% which reflects higher rate of alteration in ferricrete zone, mottle zone and sparolite, and CIA for secondary detrital laterites (Sriniketan, Birbhum) varies from 83 to 93% which signifies good preservation of chemically altered minerals.

6.3 Theories of Lateritization and Applicability in Study Area

In the investigation of laterites, one of the perennial problems is the explanation of mode of laterite formation (i.e. transfer of Fe in lateritization process), by in situ weathering, relative or absolute accumulation or by continual weathering and erosion of a long exposed landscape. The members of laterite families have different properties but similar genesis (Schellmann 1981). There have been many hypotheses or theories proposed for laterite formation, all containing elements of truth, but there still seem to be no one all-encompassing theory and perhaps there never will be (Bourman 1993). Here, a brief review of those theories is depicted to get an idea of processes of ferricrete formation and most suitable model of lateritization processes is framed for the study area, as well as for *Rarh* Bengal.

6.3.1 Swanson's Hypothesis

According to Swanson (1923), the chemical process of lateritization means the elimination of the alkalies, alkaline earths, and silica from the original rock, and the persistence of the hydrated oxides of aluminium and iron with a small amount of titanium oxide. It appears that the process of lateritization has more to do with the character of the laterite formed than the nature of the parent rock. The laterites are restricted to the regions of tropical heat and heavy rainfall, wet–dry season, luxuriant vegetation and elevated plains or plateau, having gently sloping land surface (Swanson 1923). The rocks, in which plagioclase feldspar are abundant with their usual concomitants of ferromagnesian minerals, form laterites by decomposition in situ. The silicates are decomposed to clays, and free silica is removed from the profile in solution. In the zone of concretion, the leaching is carried to finish and it is characterized by the migration of iron to the surface accompanied by the separation of Fe and Al hydroxides (Swanson 1923; Ghosh and Guchhait 2019).

6.3.2 Residuum Theory

The theory prefunded by McFarlane (1976) that laterite was the residuum of extreme weathering and differential removal of material by chemical solution became widely accepted (McFarlane 1976). He presented a composite geomorphic model of laterite formation, considering the positive aspects of the land surface reduction theory of Trendall (1962) and de Swardt (1964). In that model, original iron precipitates were believed to have formed in the narrow fluctuating range of a groundwater table, which fell as the land surface was lowered due to erosion. With the cessation of down wasting and stabilization of the water table, the residuum was thought to have been derived from the underlying pallid zone which was developed by leaching through the permeable in situ laterite after uplift and incision of laterite (McFarlane 1976; Ghosh and Guchhait 2019). That groundwater laterite was thought to have developed as a mechanical residuum.

6.3.3 Chemical Diffusion

Mann and Ollier (1985) suggested that the upward chemical diffusion of ferrous iron in solution from a bedrock weathering front, through deeply weathered profiles, followed by oxidation and precipitation of ferric oxyhydroxides at the water table, could account for occurrences of ferruginous duricrust above tens to hundreds of metres of pallid zone clay. This model implies that the development of both the pallid zone and the ferricrete by the one integrated process. This model neglects the role of lateral accession of iron oxides, groundwater fluctuation, and topography (Ghosh and Guchhait 2019).

6.3.4 In Situ Lateritization Model

Nahon et al. (1977), Tardy and Nahon (1985) and Tardy (1992) favoured in situ development of laterite profile (Fig. 6.2). The lower horizons preserved bedrock texture and weathering was considered to have been so voluminous. The removal of iron around voids, canalicules, and channels from a fine saprolite which became white, called pallid zone. The mottled zone formed with the absolute accumulation of iron in kaolinized bedrock involving the epigenetic replacement of kaolinite by haematite. Soft nodular and hard nodular iron crusts formed due to the transformation of soft yellow plasma into pisoliths. The hard nodular iron crust was characterized by an increase in haematite content. Goethite rinds occur on haematite nodules in the upper few centimetres if the duricrust. According to this model, a laterite profile

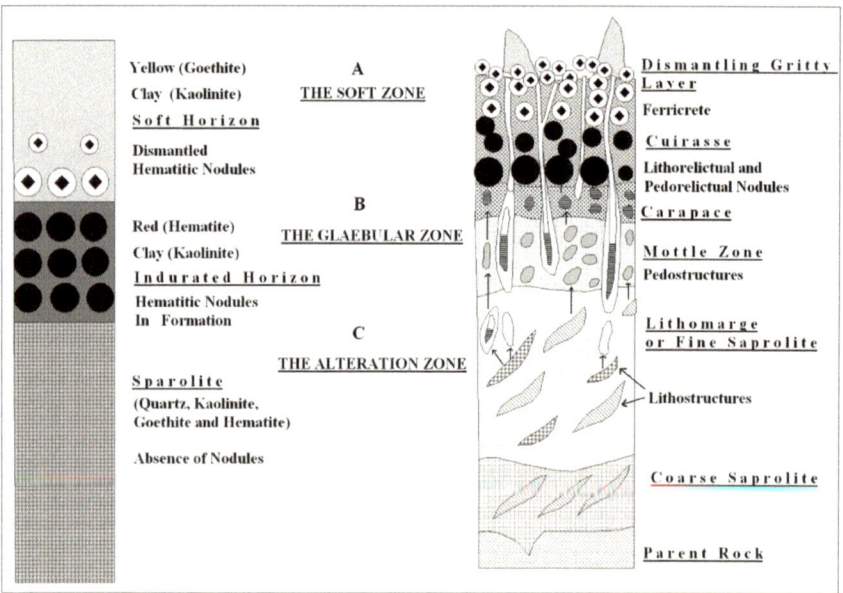

Fig. 6.2 An applicable schematic model (modified from Tardy 1992) depicting in situ development of laterite profile with successive formation of coarse saprolite, lithomarge, mottle zone, ferricrete, and gritty layer due to deep basal weathering of parent rock (similar signatures of lateritization found in the sections of Baramasia, Nalhati, and Icchanagar (Ghosh and Guchhait 2019)

has three distinct board zones—(1) zone of alteration at the base (saprolite), (2) a glaebular zone at the middle part, and (3) a soft zone, non-indurated at the upper part of profile. In the primary laterites of Indian plateaus, this in situ model fits in many cases (Row Chowdhury et al. 1965; Roy Chowdhury 1986; Ghosh and Guchhait 2019).

6.3.5 Continual Weathering Model

This model, proposed by Bourman (1993), revealed that ongoing weathering under favourable tropical climate may be interrupted by tectonic activity, burial by terrestrial sediment or submergence by a lake or the sea. Within the zone of water table fluctuation, primary iron minerals within the basement rocks were degraded by weathering under reducing conditions, forming ferrous iron that was redistributed and segregated within the weathered and partly kaolinized rock to form ferric-iron-rich mottles under oxidizing conditions (Bourman 1993). As surface weathering and erosion proceeded, the iron segregations, largely as haematite mottles, were progressively exposed at the surface, where they hardened and formed ferricrete. The pallid zones may continue to develop after uplift and that vermiform ferricrete may develop on very subdued relief by the modification of former ferricrete (Ghosh and Guchhait 2019).

6.3.6 Applicable Theory of Lateritization in Study Area

In the *Rarh* Plain, the processes of primary and secondary laterite formation are slightly different from the above-mentioned processes on the basis of magnitude of involving factors (i.e. type of weathered materials, source of ferralitic materials, wet–dry types of tropical climate, fluctuation of groundwater table, topographic positions and erosion, stability of favourable environment etc.). In the tropical geoclimatic settings, the processes of lateritization (transfers of iron), latosolization (residual accumulation of iron), desilication (loss of silica from the profile), and rubification (reddening the regolith and soil horizons with iron oxides) were simultaneously operated to develop the laterite profiles (Schaetzl and Anderson 2005; Ghosh and Guchhait 2019). Based on the field observations and above-mentioned theories or hypotheses or models, five following necessary phases of lateritization are identified in the formation of in situ and ex situ modes of laterites (Fig. 6.3).

(a) *Movement of Iron and Silica in the Profile*

There are three accepted modes of iron enrichment: vertical leaching, capillary rise, and fluctuation of water table. It is proposed that lateral migration of Fe and Al involves the accumulation at preferred sites. The change to oxidizing conditions above water table would lead to the oxidation and precipitation of ferrous ion

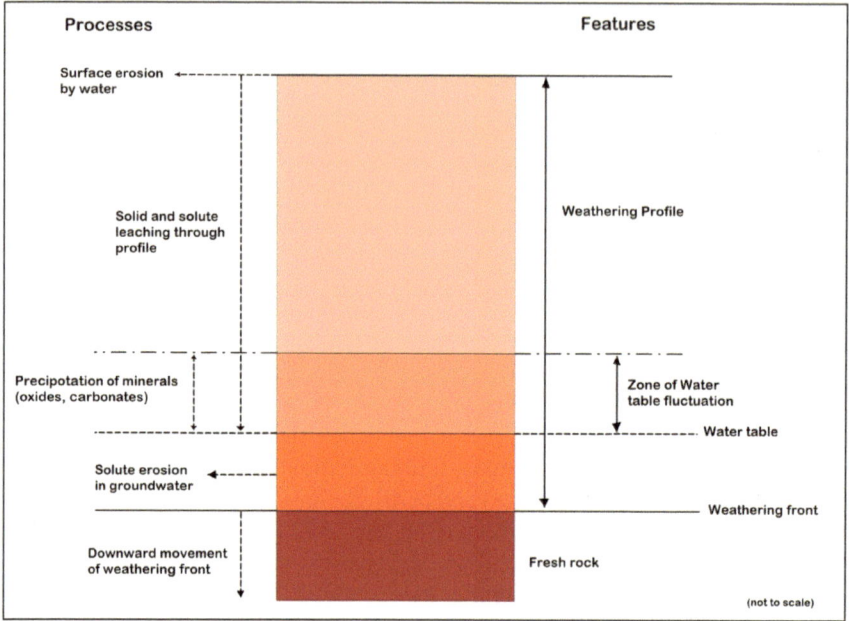

Fig. 6.3 A schematic diagram illustrating the dynamics of weathering products in a laterite profile and the controlling processes (modified from Taylor and Eggleton 2001)

(Bourman 1993). The high content of Fe_2O_3 (26.2%) and Al_2O_3 (36.71%) in the ferricrete of Nalhati profile signifies a greater upward translocation of ferruginous materials at the duricrust and low content of SiO_2 (9.76%) reflects intensive leaching of silica from the top of profile. Primary silicates are kaolinized in the early stage of weathering and most of the alkalis and alkaline materials are removed through vertical leaching during the wet phase. Water does not enter between the successive layers (1:1 clay layer) of kaolinite crystals, so the kaolinite is non-swelling (Tardy 1992). The occurrence of lithomarge kaolinite between ferricrete and bedrock acts as a perched aquifer, influencing the loss of water and free silica through upper slope of kaolinite zone (Ghosh and Guchhait 2019).

(b) *Wet–Dry Period and Groundwater Fluctuation*

Alternate wetting and drying with the fluctuation of groundwater table, most possibly accompanied by local reduction and re-oxidation, is believed to be the main cause of the movement and re-crystallization of iron (Ghosh and Guchhait 2019). The field study in Nalhati, Bhatina, and Pansiuri sections reveals that the fluctuation of groundwater table (i.e. dug well data) varies from 8 to 11 m in between rainy and dry season. So, the wet period influences leaching of silica from the profile and the dry period favours significant dehydration of Fe–Al oxides including centripetal accumulation of haematite nodules. Within the zone of water table fluctuation, primary iron minerals within weathered basement rocks are degraded by weathering under reduction condition, forming ferrous iron which is redistributed and segregated in lithomarge to form ferric-iron-rich mottles in oxidizing condition (Bournman 1993).

(c) *Role of Ferruginous Hydrogels*

Pascoe (1964) and Roy Chowdhury et al. (1965) emphasized the role of sols and gels in the formation of laterite. It is suggested that the alumina-rich layer acts as a semi-permeable membrane preventing the movement of iron oxides downwards and colloidal silica upwards. In wet phase, the massive ferricrete of in situ laterite profile acts as semi-permeable membrane of Fe–Al sols which only permits leaching of silica and kaolin through channels or tubes from the mottled zone. The precipitating hydrogels of ferrous and alumina are dehydrated later due to alternate wet and dry seasons. The suspended ferruginous matters (gels) have been deposited at the outer base of the laterite profile by spring discharge water and basal sapping, as complicated red bands of laminated lithomarge and ferruginous grit (Pascoe 1964; Ghosh and Guchhait 2019).

(d) *Centripetal Accumulation of Fe–Al Oxides*

Nahon (1986) and Tardy (1992) described the metabolism of ferricrete–concretion of iron nodules designates the mechanism of cementation and induration by centripetal accumulation of material in pores of small size (Fig. 6.4a). An iron crust is built at the top of profile by a combination of successive small-scale migration of iron, dissolution of kaolinite and quartz grains, formation of voids, secondary accumulation of kaolinite and ferruginization of these accumulations. In a sequence of ferricrete development from mottles to sub-nodules, nodules and to meta-nodules, iron content is increased and quartz content is decreased drastically. In the first phase, goethite, a hydrated mineral, prevails during the wet season, while in second phase haematite, a dehydrated mineral, dominates during dry season (Fig. 6.4b).

(e) *Lateral Accumulation of Iron*

During continuous erosion, the transfer of iron from high-level primary laterites to downslope low-level primary or ex situ laterites is possible through the lateral accumulation or groundwater suspension (as ferruginous hydrogels are insoluble) or mechanical contribution of Fe–Al oxides or nodules (Fig. 6.5). During slope retreat, the slope deposition and pediment formation are occurred by the down-wasted ferruginous materials from the upslope primary ferricrete. At the plateau margins or edge of slope, slabby ferricrete develops where laterally moving perched groundwater approaches the surface and precipitated iron oxides in near-horizontal layer (Bourman 1993). It is generally agreed that in this situation, the laterite is partly detrital, but in later, the lateritization process or duricrust is in situ to the profile (Ghosh and Guchhait 2019).

Fig. 6.4 a An applicable
model depicting formation of
macrovoids with kaolinite
and mottles in laterite from
bleaching of primary
lithostructure (modified from
Tardy et al. 1991), and **b** a
possible model depicting
successive formations of
lithomarge, macrovoids,
Fe-mottles, haematite
nodules and ferricrete in the
study area (modified from
Nahon 1986; Tardy et al.
1991; Tardy 1982; Ghosh
and Guchhait 2019)

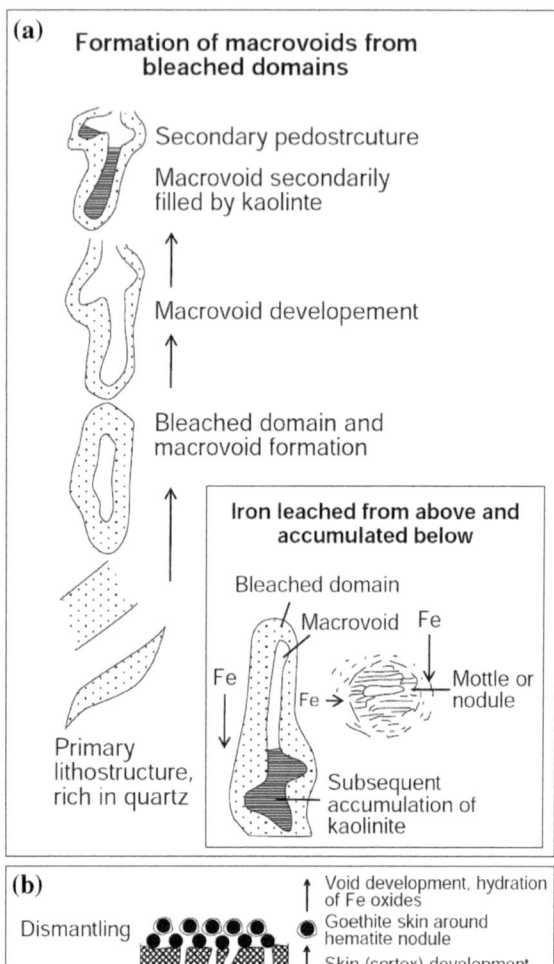

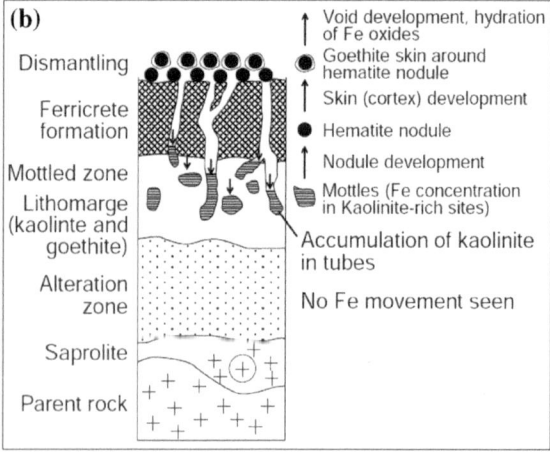

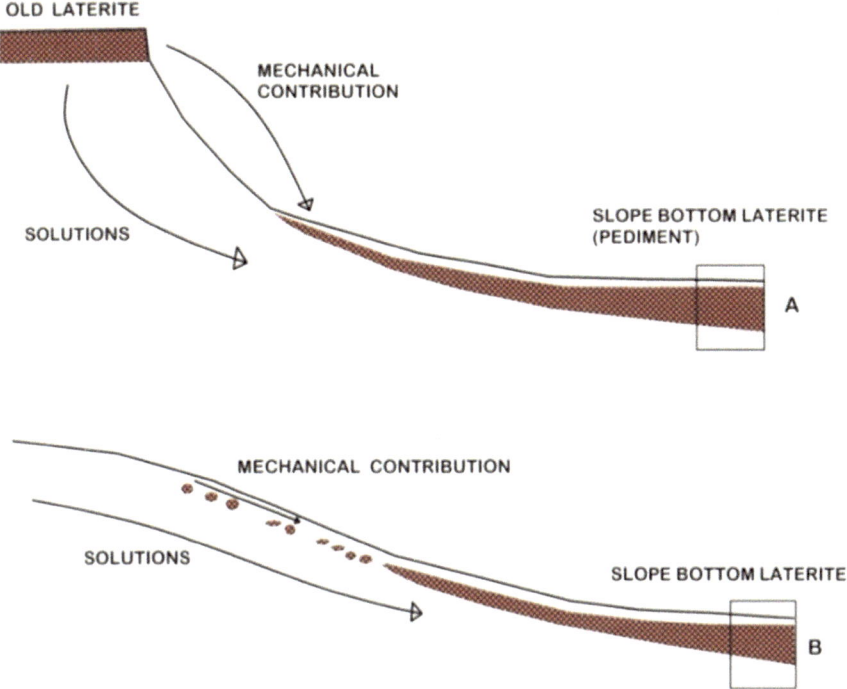

Fig. 6.5 Laterite development by absolute accumulation–slope bottom laterites receives lateral contribution of ferruginous materials from higher topographic positions, and **a** is reputedly 'secondary' and **b** 'primary' (McFarlane 1976; Ghosh and Guchhait 2019)

References

Birkeland PW (1984) Soils and geomorphology. Oxford University Press, New York

Bourman RP (1993) Perennial problems in the study of laterite: a review. Aust J Earth Sci 40(4):387–401

de Swardt AMJ (1964) Lateritization and landscape development in parts of equatorial Africa. Zeitschrift fur Geomorphologie 8:313–333

Ghosh S, Guchhait SK (2019) Modes of formation, Palaeogene to Early Quarternary Palaeogenesis and geochronology of laterites in Rajmahal Basalt Traps and Rarh Bengal of Lower Ganga Basin. In: Das BC, Ghosh S, Islam A (eds) Quaternary geomorphology in India. Springer, Singapore, pp 25–60

Mann AW, Ollier CD (1985) Chemical diffusion and ferricrete formation. Catena Suppl 6:151–157

McFarlane MJ (1976) Laterite and landscape. Academic Press, London

Meshram RR, Randive KR (2011) Geochemical study of laterites of the Jamnagar district, Gujarat, India: implications on parent rock, mineralogy and tectonics. J Asian Earth Sci 42:1271–1287

Nahon D (1986) Evolution of iron crusts in tropical landscapes. In: Colman SM, Dethier DP (eds) Rates of chemical weathering of rocks and minerals. Academic Press, Orlando, pp 169–191

Nahon D, Janot C, Karpoff AM, Paquet H, Tardy Y (1977) Mineralogy, petrography and structures of iron-crusts developed on sandstones in the western part of Senegal. Geoderma 19:263–277

Pain CF, Ollier CD (1996) Regolith stratigraphy: principles and problems. AGSO J Aust Geol Geophy 16(3):197–202

Pascoe EH (1964) A manual of the geology of India and Burma, vol 3. Geological Survey of India, Delhi

Roy Chowdhury MK (1986) Concepts on the origin of Indian laterite in historical perspective. Proc Indian Natl Sci Acad A 52(6):1307–1323

Roy Chowdhury MK, Venkatesh V, Anandalwar MA, Paul DK (1965) Recent concepts on the origin of Indian laterite. Proc Natl Acad Sci India Sect A: Phys Sci A 31(6):547–558

Schaetzl RJ, Anderson S (2005) Soils: genesis and geomorphology. Cambridge University Press, Cambridge

Schellmann W (1981) Considerations on the definition and classification of laterites. In: Proceedings of the international seminar on lateritisation processes, Trivandum, India

Schellmann W (1986) A new definition of laterite. In: Banerjee PK (ed) Lateritisation processes, vol 120. Geological Survey of India Memoir, pp 11–17

Swanson CO (1923) The origin, distribution and composition of laterite. J Am Ceram Soc 6(12). https://doi.org/10.1111/j.1151-2916.1923.tb17709.x

Tardy Y (1982) Kaolinite and smectite stability in weathering conditions. Estud Geol 38(3):295–312

Tardy Y (1992) Diversity and terminology of laterite profile. In: Martini IP, Chesworth W (eds) Weathering, soils and paleosols. Elsevier, Amsterdam, pp 379–405

Tardy Y, Nahon D (1985) Geochemistry of laterites, stability of Al-goethite, Al-hematite and Fe^{3+}-kaolinite in bauxites and ferricretes: an approach to the mechanism of concretion formation. Am J Sci 285:865–903

Tardy Y, Kobilsex B, Paquet H (1991) Mineralogical composition of geographical distribution of African and Brazilian peri-Atlantic laterites: the influence of continental drift and tropical paleoclimates during the past 150 million years and implications for India and Australia. J Afr Earth Sci 12(1–2):283–295

Taylor G, Eggleton RA (2001) Regolith geology and geomorphology. Wiley, Chichester

Trendall AF (1962) The formation of apparent peneplians by a process of combined lateritization and surface wash. Zeitschrift fur Geomorphologie 6:183–197

Chapter 7
Geochronology of Laterites

Abstract There are marked similarities in between laterites of India and Australia because these continents have moved from polar to tropical palaeolatitudes over the past 80 million years. On the basis of palaeolatitudes and laterite magnetizations, the favourable optimum climate of laterite formation was prevailed in India from Late Cretaceous to Late Tertiary. Consequently, the laterites in the northern peninsula should be older than those in the south as per drifting of the Indian plate. In this chapter, the age determination, span of lateritization event, and dating data analysis are included to draw significant information about the geochronology of laterites.

Keywords Palaeolatitude · Palaeomagnetism · Cryptomelane · OSL dating · Geochronology

7.1 Span of Lateritization Event

The earliest known laterites are those of the 2200 Ma profile near Sishen, South Africa, but bauxites extend back to at least 3500 Ma, near Taldan in the Aldan Shield of Siberia (Retallack 2010). The earliest laterites of India were dated back to Early Palaeocene–Early Oligocene, found in Gujarat and Thar Desert (Sychanthavong and Patel 1987; Meshram and Randive 2011; Ghosh and Guchhait 2019). The peak period of lateritization event was started in Neogene when the Indian plate was well established in tropical latitudes. According to the reconstruction of palaeolatitudes, it is found that southern India spent a longer time in the equatorial zone, i.e. between 53 million years and <32 million years (Fig. 7.1). The accelerated northward drift into Koppen's 'A' zone between 65 and 53 Ma propelled India quite rapidly into the favourable zone of laterite formation (Kumar 1986; Ghosh and Guchhait 2019). The age of lateritic weathering has also been roughly estimated on the basis of palaeomagnetic properties of iron oxides formed in laterites around 16° N (Schmidt et al. 1983). There are marked similarities in between the laterites of India and Australia because these continents have move from polar to tropical palaeolatitudes

This chapter is reproduced in part from Ghosh and Guchhait (2019).

© The Author(s), under exclusive license to Springer Nature Switzerland AG 2020   95
S. Ghosh and S. K. Guchhait, *Laterites of the Bengal Basin*,
SpringerBriefs in Geography, https://doi.org/10.1007/978-3-030-22937-5_7

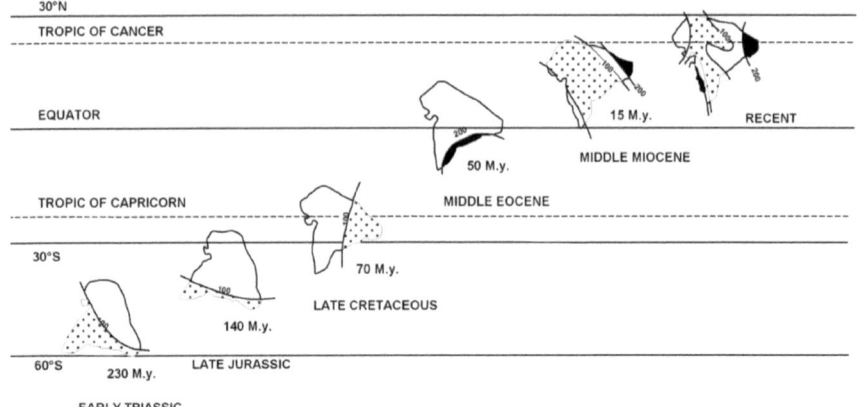

Fig. 7.1 Palaeogeographic reconstruction of the Indian plate and its entry to the region of tropics since Early Triassic. Onset of lateritization process was started in Middle Eocene. Numbers show the relative values of precipitation through geological times and no units are implied (modified from Tardy et al. 1991; Alam et al. 2003; Ghosh 2014; Ghosh and Guchhait 2019)

over the past 80 million years (Schmidt et al. 1983; Tardy et al. 1991; Retallack 2010). The optimum lateritization event was probably related to Late Oligocene to Early Miocene weathering event in India and Australia (Schmidt et al. 1976; Ollier 1988; Bourman 1993; Bird and Chivas 1993; Ghosh and Guchhait 2019).

On the basis of pole positions and comparisons with the apparent polar wander paths, Schmidt et al. (1983) have concluded that the ages of Indian laterites are grouped as follows: Late Tertiary, Mid-Tertiary, Early Tertiary, or Early Tertiary–Late Cretaceous. On the basis of palaeolatitudes and laterite magnetizations (Chemical Remnant Magnetization in haematite), the favourable optimum climate of laterite formation was prevailed in India from Late Cretaceous to Late Neogene (Kumar 1986; Ghosh and Guchhait 2019). Recently, the results of $^{40}Ar/^{39}Ar$ dating of laterite samples imply that Southern India was weathered to form laterite duricrust between ~36 and 26 Ma (Late Eocene–Oligocene) and may have been dissected mostly in Neogene (Bonnet et al. 2014; Ghosh and Guchhait 2019). The plateau top laterites of Deccan Trap is dated back to Eocene–Miocene and detrital low-level laterites with gravels and Acheulian artefacts indicate severe erosion of primary high-level laterite cover during Early Pleistocene (Fig. 7.2) (Mishra et al. 2007).

7.2 Age of Rajmahal Laterites

The post-succession of Rajmahal Basalt Traps (RBT, adjoining to study area) is in situ laterites, gravel deposits and ferruginous sandstones on the stable shelf condition of Bengal Basin. In the Bengal Basin, the laterites over Rajmahal Basalts are the

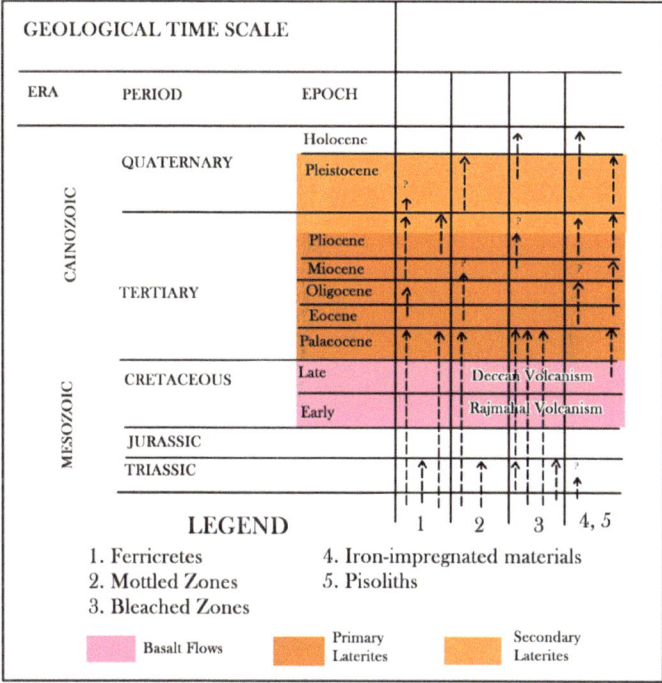

Fig. 7.2 Geological clock of laterite evolution in Indian Peninsula and the study area, showing three phase major events since Cretaceous—(1) widespread volcanism in Cretaceous Period, (2) Paleocene to Mid-Pliocene in situ intense lateritization, and (3) Pliocene to Late Pleistocene ex situ re-lateritization (Bourman 1993; Ghosh and Guchhait 2019)

oldest tropical weathering profiles. The $^{40}Ar/^{39}Ar$ geochronological ages of Rajmahal basalts and basalts of Bengal Basin were dated back at ~118 Ma (Kent et al. 2002). So, it is logical to establish that the development of in situ primary laterites on the Rajmahal basalts, Gondwana Formations and Archaean granite–gneiss is post-Cretaceous formation. One interesting observation is the Palaeogene age assigned to laterites from the Deccan Traps which are Late Cretaceous in age. Similarly, the high-level primary laterites are found on the Early Cretaceous Rajmahal Basalt Traps (situated at north-western of study area, 24° 15′–24° 50′ N, 87° 15′–87° 45′ E) and the latitudinal position of this province is quite north of Deccan Traps (between 16°–24° N, 73°–80° E) (Ghosh and Guchhait 2019). So in respect of equatorial drift of Indian plate (Late Cretaceous–Early Palaeogene), this geologic unit of Rajmahal was getting early favourable climate for lateritization. The time taken for the entry of Indian plate into the zone between 30° N and 30° S is about 65–75 million years from the present (Kumar 1986). Therefore, the Rajmahal Basalts Traps (fringing north-west of *Rarh* Bengal) were must undergone through intensive lateritization after Late Cretaceous.

In India, firstly, Sankaran et al. (1985) have applied TL dating for the age estimation of laterites and found that the age varies from 2.15 to 3.58 × 10^5 years. Cryptomelane [K_x Mn^{IV}_{8-X} Mn^{III}_x O_{16}, n H_2O] is a major Mn-oxide of many lateritic Mn-ore deposits in south India (Bonnet et al. 2015a; Ghosh and Guchhait 2019). This mineral is a suitable chronometer for of $^{40}Ar/^{39}Ar$ step heating dating of weathering processes in the lateritic soils including duricrusts and manganiferous ore deposits (Bonnet et al. 2014, 2015a, b; Beauvais et al. 2016). It has been learned that Peninsular India experienced six major phases of lateritic weathering, viz. ~53–50 Ma, ~40–32 Ma and ~30–23 Ma in the plateau tops and ~47–45 Ma, ~24–19 Ma and ~9 Ma in lowlands, pediments, and valleys (Bonnet et al. 2015a, b; Beauvais et al. 2016). The results of $^{40}Ar/^{39}Ar$ dating of laterite samples (Bonnet et al. 2014) and other dating information (Schmidt et al. 1983; Kumar 1986; Sychanthavong and Patel 1987; Tardy et al. 1991; Bourman 1993; Widdowson and Cox 1996; Rajaguru et al. 2004; Mishra et al. 2007; Retallack 2010) imply that basalts of RBT were weathered intensively to form in situ ferricrete in between ~36 and 26 Ma (Late Eocene–Oligocene) and may had been dissected mostly since Neogene under favourable lateritization climate (becoming source materials of ex situ secondary laterites) (Ghosh and Guchhait 2019).

7.3 Key Features and Age of Secondary Laterites

Niyogi et al. (1970) assigned Neogene and Early Pleistocene age for the lithomargic kaolinite clay and laterites, respectively. The ferruginous soils of the present study area are regarded as the oldest soils from the Indian part of these plains (Singh et al. 1998). Niyogi (1975), Vaidyanadhan and Ghosh (1993) and Singh et al. (1998) have been assigned ages of 305–1000 ka, i.e. Early–Middle Pleistocene, reflecting autochthonous origin of ferruginous soils. The archaeological evidence suggest an Early-to-Middle Pleistocene origin of secondary detrital laterites (Rajaguru et al. 2004). The upper dismantled surface and ex situ laterites have bears the evidence of Acheulian (Early–Middle Pleistocene) to Mesolithic (40–150 ka) artefacts in India as well as *Rarh Bengal* (mainly in the districts of Birbhum, Bardhaman, and Bankura).

For accurate chrono-stratigraphic age detection, sampling for optically stimulated luminescence (OSL) dating of sample laterite lithofacies in Sriniketan of Birbhum, Bishnupur of Bankura, and Garhbeta of West Medinipur have been done by Chakraborti (2011) (Table 3.6). In Garhbeta section, the samples have been taken from the hard crust (1.6 m depth), pebble horizon (2.5 m depth), loose murram layer (3.2 m depth), and siltstone unit (5.2 m depth). For such investigation, samples have been collected from the upper three horizons of lateritized sediments including hard crust (0.3 m depth), ferruginized coarse sand–pebble horizon (1.9 m depth) and finely laminated siltstone (2.9 m depth) from Sriniketan section. Similarly, numerous samples have also been collected from three layers—newer alluvium (0.9 m depth), older alluvium (2.2 m depth), and *murram* layer (3.4 m depth) from Bishnupur section (Fig. 7.3a) (Ghosh and Guchhait 2019).

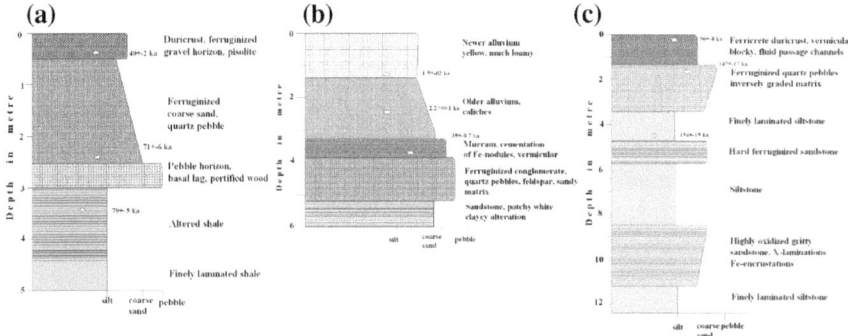

Fig. 7.3 **a** Formation of ex situ ferruginized duricrust with gravel horizons on finely laminated shale in Sriniketan, Birbhum, **b** successive lithofacies of newer alluvium, older alluvium and vermicular ferruginous hard crust in Bishnupur, Bankura, and **c** development of Early Quaternary ferricrete and ferruginized quartz pebbles on finely laminated alternate siltstone and sandstone sequences in Garhbeta, West Medinipur (age in ka derived from OSL dating of samples) (Ghosh and Guchhait 2019)

7.3.1 Sriniketan Lithosection

The ex situ laterite profile of Sriniketan, Birbhum (23° 41′ 31″ N and 87° 40′ 31″ E) is characterized by (1) pebble horizon (2.6–3.0 m depth), (2) ferruginized coarse sand (0.55–2.6 m depth) and (3) duricrust (up to 0.55 m depth) (Fig. 7.3a). Pebble horizon is characterized by lag deposits, constituting of pisoids, quartz pebbles and petrified woods of varying sizes set in a ferruginized matrix of sands. The thick layer of coarse sands constitutes gravels and pebbles of varying sizes and ferricrete pisolites which may be derived from distant locations of primary laterites. The thickness of duricrust generally varies from 30 to 50 cm and many a places absence of this layer are very common. This duricrust is nothing but a highly ferruginized or iron-cemented gravel and pebble horizon, constituting of quartz, ferricrete pisolite, petrified wood fragment and altered as well as fresh feldspar clasts. Basically, it appears as a conglomerate ferricrete and Gmg (i.e. inverse to normal grading matrix-supported gravels) fluvial facies.

7.3.2 Bishnupur Lithosection

In the exposed profile of Bishnupur, Bankura (23° 05′ 28″ N and 87° 16′ 15″ E), one conglomerate ferruginous unit un-conformably overlies the grey-coloured coarse sandstone unit (Fig. 7.3b). This conglomerate unit with murram at its upper part is overlain successively by older alluvium and newer alluvium sediments. The lowermost sandstone unit (5.2–6.0 m depth) is relatively more indurated and lithified than upper older and newer alluviums. It is relatively coarse-grained and at places granular, yellowish-white in colour and characterized by white clayey alteration. This Paleogene sandstone unit is un-conformably overlain by a ferruginized conglomerate unit (3.8–5.1 m depth) which is oligomictic in character with pebbles of dominant quartz and little feldspar. The matrix or groundmass is highly cemented by ferruginous cement (mainly limonite and goethite) which makes it very indurated. So, it can be identified as modified ferricrete–duricrust of *Rarh* Plain. The upper part of this conglomeratic ferricrete duricrust (3.2–3.8 m depth) is represented by murram with iron-stained quartz pebbles of 1–2 cm in size and ferricrete nodules in a ferruginized sand-sized matrix (*Gmg* facies).

7.3.3   Garhbeta Lithosection

The right bank of Silai River, exposes a vertical profile of almost 12- to 14-m-thick package of alternate sandstone–siltstone sequence with overlying ferruginized sequence of fining-upward fluvial sediments at Garhbeta of West Medinipur district (Fig. 7.3c). The exposed vertical section of about 12.5 m may be categorized into two geological units with distinctive characteristics. The lower lithofacies (3.5–12.5 m depth) is mainly constituted of Paleogene to Neogene alternate sandstone–siltstone lithofacies whose contacts at several levels are marked by thin but hard ferruginized sandstone layers of 4–8 cm thick. The lowermost siltstone lithofacies are very finely laminated yellowish-white and purple in colour and at places show whitish patchy clayey alteration. The ferruginized sandstone unit is generally grey in colour and towards the top portion of the sequence reddish in colour, coarse-grained gritty in character and at places contain few quartz pebbles. It is found that the sandstone unit (8.5–11.7 m depth) is reddish in colour, highly oxidized and characterized by high degree of iron encrustation features. The top laterite is characterized by the topmost ferricrete duricrust followed by murram zone and lowermost quartz pebble horizon. This litho unit is constituted mainly of finer quartz grains and cemented by goethite and limonite. This layer of duricrust with pebble horizon (1.5–3.3 m depth) un-conformably overlies the Neogene siltstone unit (3.3–4.8 m depth).

7.3.4   OSL Dating

For accurate chrono-stratigraphic age detection, sampling for optically stimulated luminescence (OSL) dating of sample laterite, lithofacies in Sriniketan of Birbhum, Bishnupur of Bankura, and Garhbeta of West Medinipur have been done. In Garhbeta section, the samples have been taken from the hard crust (1.6 m depth), pebble horizon (2.5 m depth), loose murram layer (3.2 m depth) and siltstone unit (5.2 m depth). Sampling has been done from the upper three horizons of lateritized sediments including hard crust (0.3 m depth), ferruginized coarse sand–pebble horizon (1.9 m depth) and finely laminated siltstone (2.9 m depth) at Sriniketan section. Similarly, numerous

Table 7.1 Summary of U, Th, and K elemental concentrations, annual dose rate, equivalent dose and optical ages of ex situ laterite samples from the study area

Sl No.	Sample No.	Sample horizon	Depth in m	No. of discs	U (ppm)	Th (ppm)	K %	Equivalent dose (avg) in grey	Dose rate grey/ka	Age (avg) ka
Sriniketan section										
1	OSLD 1	Hard crust	0.4	45	1.94	11.93	0.88	78 − 2	1.9 ± 0.09	40 ± 2
2	OSLD 2	Ferruginized sand-stone–pebble	2.5	45	1.17	8.56	1.63	153 − 9	1.1 ± 0.01	71 ± 6
3	OSLD 3	Siltstone	3.4	41	1.71	14.17	1.63	203 − 9	2.5 ± 0.01	79 ± 5
Bishnupur section										
4	OSLD 4	Younger alluvium	1.6	12	2.95	17.04	0.95	35 + 0.4	2.33 ± 0.03	1.5 ± 0.2
5	OSLD 5	Older alluvium	2.5	12	3.81	18.17	1.18	6.2 + 0.5	2.8 ± 0.01	2.2 ± 0.1
6	OSLD 6	Murram	3.9	42	3.56	37.59	0.65	12 + 2	3.4 ± 0.02	35 ± 0.7
Garhbeta section										
7	OSLD 7	Hard crust	0.3	48	4.34	16.42	0.43	214 − 14	2.23 ± 0.01	96 ± 8
8	OSLD 8	Pebble horizon	1.9	45	1.61	11.44	0.4	205 − 21	1.4 ± 0.08	147 ± 17
9	OSLD 9	Siltstone	4.8	36	1.63	13.71	0.68	264 − 21	1.7 ± 0.09	154 ± 15

Source Chakraborti (2011), Ghosh and Guchhait (2019)

samples have also been collected from three layers—newer alluvium (0.9 m depth), older alluvium (2.2 m depth), and murram layer (3.4 m depth) at Bishnupur section (Table 7.1).

The possible age, as determined by OSL method for the Sriniketan section, shows that the age of sedimentation or time of cut off from the sunlight for the hard crust (0.45 m depth) is 40 ± 2 ka. The age of ferruginized sandstone–pebble horizon (with petrified wood) is about 71 ± 6 ka. The age of laminated siltstone is 79 ± 5 ka. The layer of ferruginous hard crust with gravels was probably developed in Late Pleistocene (well within ~125–10 ka BP) (Ghosh and Guchhait 2019).

The age as determined from the samples for Bishnupur section shows the age of sedimentation of the murram zone is more than 35 ± 0.7 ka as it was cut off from the sunlight at this date. Similarly, the older alluvium is older than 2.2 ± 0.1 ka age and newer alluvium is older than 1.5 ± 0.2 ka. The ferruginous crust is again assigned as age of Late Pleistocene and other above alluvium units are categorized as the lithofacies of Late Holocene to Recent (Ghosh and Guchhait 2019).

The sedimentation age for the Garhbeta section shows that the lower siltstone unit (4.8 m depth) and pebble horizon (1.9 m depth) was cut off from the sunlight before 154 ± 15 ka and 147 ± 17 ka, respectively. Interestingly, both these units are intensively ferruginized under the tropical climate. Alongside the ferruginous hard crust (0.3 m depth) was cut off from the sunlight before 96 ± 8 ka. Yet again the laterite hard crust is assigned an age of Late Pleistocene. Based on the OSL dating data, it can be said that climate for lateritization was prevailed at Middle Pleistocene and became intense in Late Pleistocene, forming lateritic hard crust (Ghosh and Guchhait 2019).

References

Alam M, Alam MM, Curray JR, Chowdhary MLR, Gandhi MR (2003) An overview of the sediment geology of the Bengal Basin in relation to the regional tectonic framework and basin-fill history. Sediment Geol 155(3–4):179–208

Beauvais A, Bonnet NJ, Chardon D, Arnaud N, Jayananda (2016) Very long-term stability of passive margin escarpment constrained by $^{40}Ar/^{39}Ar$ darting of K-Mn oxides. The Geological Society of America. https://doi.org/10.1130/g373031.1

Bird MI, Chivas AR (1993) Geomorphic and palaeoclimatic implications of an oxygen—isotope chronology for Australian deeply weathered profiles. Aust J Earth Sci 40(4):345–358

Bonnet NJ, Beauvais A, Arnaud N, Chardon D, Jayananda M (2014) First $^{40}Ar/^{39}Ar$ dating of intense Late Palaeogene lateritic weathering in peninsular India. Earth Planet Sci Lett 386:126–137

Bonnet N, Arnaud N, Beauvais A, Chardon D (2015a) Deciphering post-Deccan weathering and erosion history of south India Archean rocks from cryptomelane $^{40}Ar - ^{39}Ar$ dating. Geophys Res Abstr 17:9114

Bonnet N, Beauvais A, Chardon D, Arnaud N (2015b) Evolution of the south-west Indian continental divergent margin: constraints from $^{40}Ar - ^{39}Ar$ dating of lateritic paleolandsurfaces. Geophys Res Abstr 17:9377

Bourman RP (1993) Perennial problems in the study of laterite: a review. Aust J Earth Sci 40(4):387–401

Chakraborti S (2011) Final report on Quaternary laterites in the western districts of West Bengal—their geomorphology, stratigraphy, genesis and implications for climate change. Geological Survey of India Eastern Region, Kolkata, pp 1–88

Ghosh S (2014) Palaeogeographic significance of ferruginous gravel lithofacies in the Ajay-Damodar Interfluve, West Bengal, India. Int J Geol Earth Environ Sci 4(3):81–100

Ghosh S, Guchhait SK (2019) Modes of formation, Palaeogene to Early Quaternary Palaeogenesis and geochronology of laterites in Rajmahal Basalt Traps and Rarh Bengal of Lower Ganga Basin. In: Das BC, Ghosh S, Islam A (eds) Quaternary geomorphology in India. Springer, Singapore, pp 25–60

Kent RW, Pringle MS, Muller RD, Saunders AD, Ghose NC (2002) [40]Ar/[39]Ar geochronology of the Rajmahal Basalts, India and their relationship to the Kerguelen Plateau. J Petrol 43(7):1141–1153

Kumar A (1986) Palaeolatitudes and the age of Indian laterites. Palaeogeogr Palaeoclimatol Palaeoecol 53:231–237

Meshram RR, Randive KR (2011) Geochemical study of laterites of the Jamnagar district, Gujarat, India: implications on parent rock, mineralogy and tectonics. J Asian Earth Sci 42:1271–1287

Mishra S, Deo S, Rajaguru SN (2007) Some observations on the laterites developed on Deccan Trap: implications for the Post-Deccan Trap denudational history. J Geol Soc India 70:469–475

Niyogi D (1975) Quaternary geology of the coastal plain in West Bengal and Orissa. Indian J Earth Sci 2:51–61

Niyogi D, Mallick S, Sarkar SK (1970) A preliminary study of laterites of West Bengal, India. In: Chatterjee SP, Das Gupta SP (eds) Selected papers physical geography (vol 1). 21st international geographical congress, Calcutta, National Committee for Geography, pp 443–449

Ollier CD (1988) The regolith in Australia. Earth Sci Rev 25:355–361

Rajaguru SN, Deo SG, Mishra S, Ghate S, Naik S, Shirvalkar P (2004) Geoarchaeological significance of the detrital laterites discovery in the Karha Basin, Pune District, Maharastra. Man Environ XXIX(1):1–6

Retallack GJ (2010) Lateritization and bauxitization events. Econ Geol 105:655–667

Sankaran AV, Nambi KSV, Sunta CM (1985) Thermoluminescence of laterites: applicability in dating. Nucl Tracks 17(5):177–183

Schmidt PW, Currey Ollier CD (1976) Sub-basaltic weathering, damsites, palaeomagnetism and the age of lateritization. J Geol Soc Aust 23(4):367–370

Schmidt PW, Prasad V, Raman PK (1983) Magnetic ages of some Indian laterites. Palaeogeogr Palaeoclimatol Palaeoecol 44:185–202

Singh LP, Parkash B, Singhvi AK (1998) Evolution of the Lower Gangetic Plain landforms and soils in West Bengal, India. CATENA 33:75–104

Sychanthavong SPH, Patel PK (1987) Laterites and lignites of northwestern India and their relevance to the drift tectonics of the Indian Plate. Curr Sci 56(10):469–473

Tardy Y, Kobilsex B, Paquet H (1991) Mineralogical composition of geographical distribution of African and Brazilian peri-Atlantic laterites: the influence of continental drift and tropical paleoclimates during the past 150 million years and implications for India and Australia. J Afr Earth Sci 12(1–2):283–295

Vaidyanadhan R, Ghosh RN (1993) Quaternary of the east coast of India. Curr Sci 31(6):231–232

Widdowson M, Cox KG (1996) Uplift and erosional history of the Deccan Traps, India: evidence from laterites and drainage patterns of the Western Ghats and Kankan Coast. Earth Planet Sci Lett 137:57–69

Chapter 8
Palaeogeographic Significance of Laterites

Abstract Many researchers have reliability on the laterites and weathered zones as palaeoclimatic indicators and as morpho-stratigraphic markers, because that ferruginous facies were the products or regolith of past weathering processes under suitable geo-climatic conditions which are not prevailing today. The residual laterite profiles of the Bengal Basin are the fossil type formed in past geological ages when climatic conditions were favourable for lateritization. These laterites were generally formed under an oxic atmosphere in the presence of abundant terrestrial biomass in an acidic environment, elevated atmospheric carbon dioxide, exceptional fossil wood preservation, and intense deep basal weathering of basalts, dolerite, gneiss, sandstones, and Neogene gravelly sediments of Bengal Basin. In this chapter, it is tried to unearth the palaeogenesis, palaeoclimatic implication, and tectono-geomorphic evolution of laterites in the shelf zone of Bengal Basin.

Keywords Palaeoclimate · Palaeogeomorphology · Lateritic lithofacies · Inversion of relief · Tectonics

8.1 Palaeoclimatic Inference

The lateritization reflects a special type of tropical climatic conditions which is characterized by the contrasted seasons (wet–dry), high temperature throughout the year (28–35 °C), annual average relative humidity of the air nearer to 60%, annual rainfall lower than 1700 mm and long dry seasons during which a relatively low thermodynamic activity of water and atmospheric relative humidity decreases (McFarlane 1976; Tardy et al. 1991). While bauxites and aluminium enrichment can stand at lower temperature (>22 °C) and are favoured by a higher thermodynamic activity of water and a higher relative humidity of the air (>80%) (Tardy 1992; Ghosh and Guchhait 2019).

It is now evidenced that climatic conditions were favourable for lateritization from Cretaceous to Palaeocene times for during that period, the Indian continent crossed

---

This chapter is reproduced in part from Ghosh and Guchhait (2019).

the zone between 30° S and 0° latitude (Fig. 8.1) (Schmidt et al. 1983; Kumar 1986; Tardy et al. 1991). The palaeoclimate of Eocene and Middle Oligocene was more favourable for the in situ type of laterite formation in peninsular India because in Eocene the equator was running across central Gujarat to southern West Bengal (Bardossy 1981). As part of the global drift of plates, India also had drifted (average 2–3 cm year^{-1}) from relatively colder southern latitudes (between 37° and 53° S) to the present day tropical and monsoon-dominated climatic zone, commencing from about 70 million years (Sankaran et al. 1985; Ghosh and Guchhait 2019).

The Indian plate suddenly accelerated to 20 cm year^{-1} from Late Cretaceous to Early Eocene (Chatterjee et al. 2013). The Indian Plate was rotated in an anti-clockwise direction since Palaeocene (Sychanthavong and Patel 1987), after that particular linear laterite *Rarh* belt of NE–SW direction (formerly as E–W coinciding with the alignment of perfect tropical zone for lateritization) had crossed the equatorial zone while the total rotation was 50° in between Palaeocene and Present (Fig. 8.1). At that time span, most of the primary laterites of *Rarh Bengal* (especially in the Rajmahal Basalt Traps) were developed. The laterite capping was observed over that Mio–Pliocene sandstones and it suggests continuation of favourable climate to lateritization in the post-Pliocene times (Devaraju and Khanadali 1993). The hematite (Fe_2O_3) and boehmite ($AlO(OH)$) are dehydrated minerals in secondary laterites and the presence of these two minerals refelcts a less humid and warmer tropical palaeoclimatic conditions (Early–Late Pleistocene) in the western Bengal Basin (Ghosh and Guchhait 2019). Therefore, the presence of *Rarh* laterites signifies a special warm and seasonal contrasted palaeoclimate which is not found in present decade throughout the study area. After that morpho-stratigraphic ferruginous unit, Sijua Formation with caliches (Late Pleistocene–Early Holocene, arid-semi arid phase) and Chuchura Formation (Middle Holocene–Late Holocene, sub-humid phase) were developed without any development of ferruginous facies (Ghosh and Guchhait 2019).

8.2 Palaeogeomorphic Significance

8.2.1 Geomorphic Units of Laterites

Niyogi et al. (1970) presented an organized thematic map of different lateritic lithofacies (Early Pleistocene to Late Holocene) in West Bengal in five classes (Fig. 8.2)—(1) in situ laterites on Rajmahal Basalt Trap, Archaean rocks, Gondwana sediments, and Tertiary gravelly sediments, (2) mottled laterite clay on Pleistocene upland, (3) mildly mottled lateritic clay in river valleys, (4) faintly mottled lateritic clay in Kasai–Damodar deltaic plain, and (5) mildly lateritized soil in interfluves. In most cases of *Rarh Bengal*, the extensive mottling of brown iron hydroxides and aluminium clays found in B-horizon of soils, reflecting warm–humid palaeoclimate. The possible ages of B-horizons through TL dating (Singh et al. 1998) are 6.7 ka of Bhagirathi–Ajay Plain, 5.44 ka of Ajay–Silai Plain, and 3.6 ka of Damodar Deltaic Plain, respectively (Ghosh and Guchhait 2019).

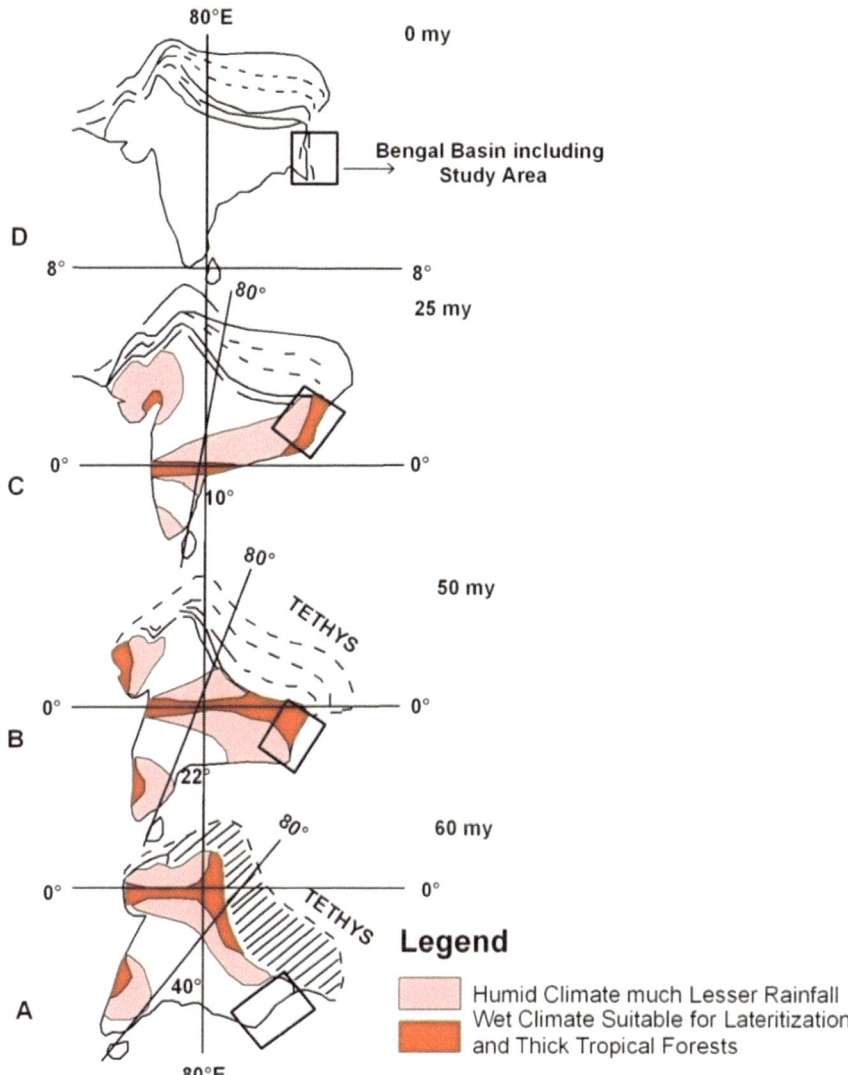

Fig. 8.1 **a, b, c,** and **d** Palaeoclimatic reconstruction of the Indian Plate (including demarcated study area) and its rotation since Early Palaeocene, during its course of drifting across the equatorial zone with the distribution of laterites (modified from Sychanthavong and Patel 1987; Ghosh and Guchhait 2019)

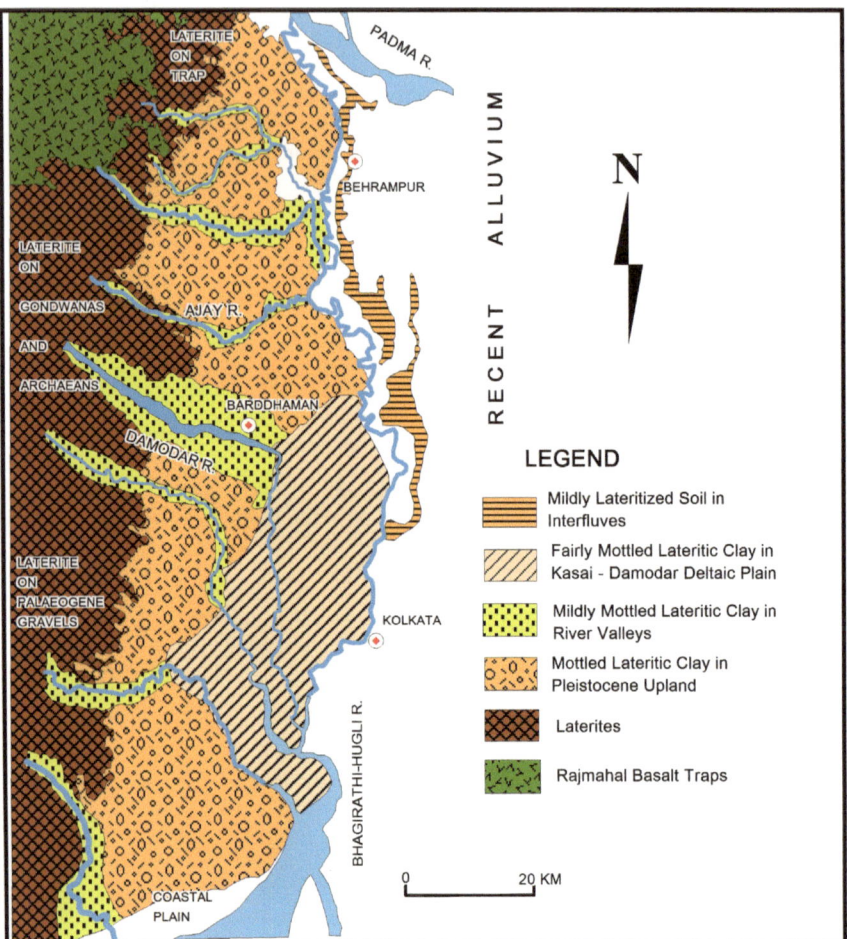

Fig. 8.2 Different litho-units of laterites and lateritic sediments in the western part of Bengal Basin, West Bengal—(1) laterites on Rajmahal Basalt Trap, Archaean rocks, Gondwana sediments, and Palaeogene gravelly sediments, (2) mottled laterite clay on Pleistocene upland, (3) mildly mottled lateritic clay in river valleys, (4) faintly mottled lateritic clay in Kasai–Damodar deltaic plain, and (5) mildly lateritized soil in interfluves (modified from Niyogi et al. 1970)

8.2.2 Inversion of Relief—Theory of Landscape Evolution

The evolution of *Rarh* laterites is directly connected with the stable shelf zone of Bengal Basin, experience maximum marine transgression, sediment depositions, tectonic uplifts, and lateritization. The dominance of kaolinite clay with presence of *hystrichospheriods* (in the pores of clay beds) indicates lacustrine to fluvio-lacustrine condition of deposition in Neogene (Mukherjee et al. 1969). The whole of the present day Bengal Basin (including Stable Shelf) was under marine water until Mioene—

Pliocene epoch and the strandline grazed the eastern margin of Peninsular Shield, i.e. much inland (towards west of study area) from the present day Orissa–Bengal coastline (Vaidyanadhan and Ghosh 1993). The stable shelf zone is separated by the Chotanagpur Foot-hill Fault (CFF) at west and the Medinipur–Farraka Fault (MFF, or called Pingla Fault) at east. Within this tectonic shelf, the *Rarh* laterites of West Bengal (NNE–SSW axis) were developed when the sea finally transgressed from this region since Late Neogene (Ghosh and Guchhait 2019). At that time, the Indian plate had been crossed the intense weathering zone of equatorial climate which was favourable for lateritization. Geomorphologically, these laterites over Paleogene–Quaternary sequences are represented by degraded badlands which are dissected by the drainage system of west to east flowing rivers of West Bengal. Only the primary laterites of Rajmahal Traps are preserved in butte type structures and under blanket of ferruginous soils. All laterites are topographically restricted within 35–115 m from mean sea level (Ghosh and Guchhait 2019).

The terrain of *Rarh* laterites is genetically linked with inversion of relief and active tectonics. Inversion of relief refers to an episode in landscape evolution when a former valley bottom becomes a ridge, bounded by newly formed valleys on each side (Pain and Ollier 1995; Ollier and Sheth 2008). Inversion of relief occurs when materials on valley floors are, or become, more resistant to erosion than the adjacent valley slopes (Pain and Ollier 1995; Ollier 1995). In the first model (Fig. 8.3), the lateral movement of water on hillsides carried weathering products from upper slopes to lower sites, when drainage was often impeded and so chemical precipitation was likely (Ollier 1995; Ollier and Sheth 2008). Gradually up to Neogene, the valley was filled with ferruginous materials and prolong lateritization formed ferricrete within Late Pleistocene. The surrounding terrain was eroded to form next valleys and the present summits or interfluves of duricrusted mesas were formed. In the

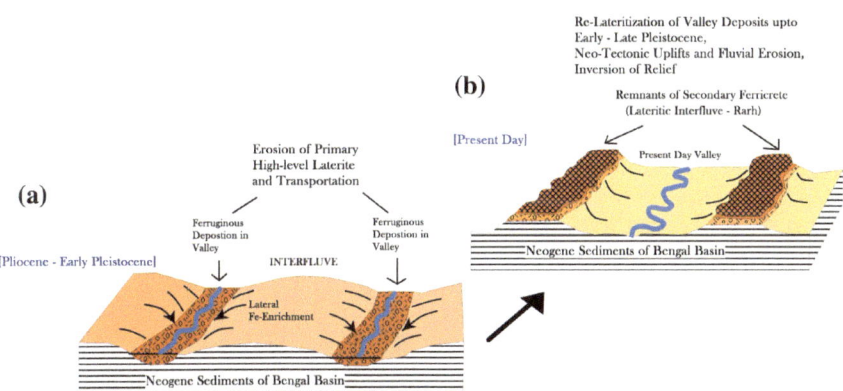

Fig. 8.3 A schematic model of secondary laterite evolution is proposed here, **a** showing the former valleys of ferruginous deposition and then re-lateritization of sediments, and **b** formation of ex situ ferricrete as present summit of mesas and present day valley, i.e. inversion of relief (modified from Pain and Ollier 1995; Ollier and Sheth 2008)

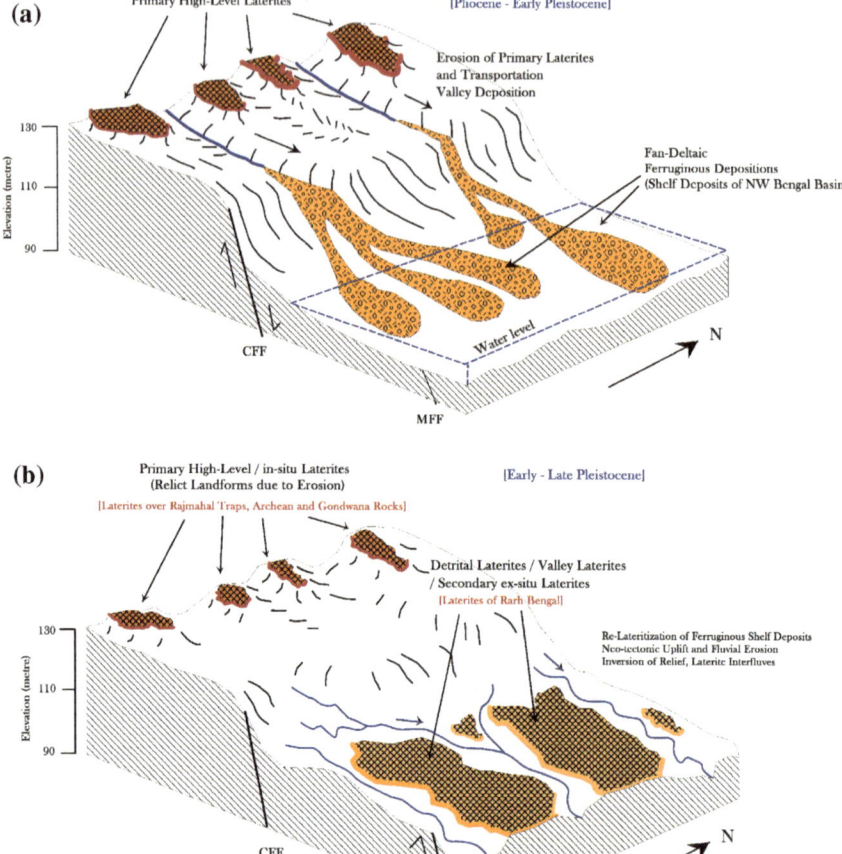

Fig. 8.4 A schematic model of *Rarh* laterite evolution, showing **a** erosion of primary laterites and ferruginous fan-deltaic depositions (modified from Mahapatra and Dana 2009) by rivers and stream in shelf zone of Bengal Basin in between Chotanagpur Foot-hill Fault (CFF) and Medinipur—Farakka Fault (MFF) up to Neogene, and **b** recession of sea, exposure of ferruginous sediments to lateritization climate (Early–Late Pleistocene), re-lateritization to form secondary *Rarh* laterites, neo-tectonic uplift, badland erosion to develop isolated summits of duricrust mesas, and inversion of relief (Ghosh and Guchhait 2019)

second model (Fig. 8.4), we reconstructed the event that up to end of Neogene the transported ferruginous materials (due to erosion of primary plateau laterites) re-deposited in the faulted Stable Shelf of Bengal Basin (under marine condition) by the drainage system of peninsular rivers as oldest fan-deltaic to para-deltaic formation in between CFF and MFF (Fig. 8.4) (Ghosh and Guchhait 2019).

8.3 Role of Tectonics

Since Early Pleistocene, the sea started regressed from that region and those valley sites of sediment deposition were subjected to further lateritization up to Middle Pleistocene and forming duricrust at top of most of ex situ laterite profiles. Additionally, there was a marine regression and uplifts in this shelf zone of Bengal Basin (i.e. *Rarh Bengal*) after the end of Miocene–Pliocene. Since Early Quaternary, the unit between MFF and CFF started to uplift (Fig. 8.5) due to re-activation of basin basement faults during occasional Himalayan upheaval and active tectonics of Bengal Basin (Ghosh and Guchhait 2015). Then during 6–7 ka, the eastern unit of Tectonic Shelf (between MFF and Damodar Fault) is subjected to subsidence (Singh et al. 1998) and the western lateritized unit (i.e. *Rarh Bengal*) is subjected to relief inversion (Ollier 1991; Pain and Ollier 1995) due to neotectonic uplifts and consecutive fluvial erosion in Holocene times. Increased precipitation during the ~15–5 ka period of peak monsoon recovery probably increased discharge and promote incision and wide spread badland formation (Sinha and Sarkar 2009). As fluvial erosion proceeds, the valley floor becomes a ridge and interfluves (i.e. laterites of *Rarh Bengal*) bounded by newly formed Late Quaternary valleys on each side (Ghosh and Guchhait 2019).

Due to episodic neo-tectonic uplifts (Fig. 8.5), the reworked laterites with gravels are found at a distance from MFF or Pingla Fault at sites of Rampurhat, Mallarpur (24° 04′ 31″ N, 87° 41′ 00″ E), Labhpur, Bolpur, Guskara (23° 29′ 17″ N, 87° 44′ 48″ E), Khandoghosh (23° 12′ 51″ N, 87° 41′ 30″ E), and Kharagpur (22° 20′ 43″ N, 87° 19′ 35″ E), following the fault-line scarp of Pingla Fault (Ghosh and Guchhait 2019). This tectonic upliftment of Stable Shelf (Singh et al. 1998) under tropical wet-dry palaeoclimate influenced the deep weathering of Fe-minerals, seasonal groundwater regime, sub-aerial exposure of valley sediments, intense leaching of silica to a depth, irreversible dehydration of Fe–Al oxides, well subsurface drainage and post-lateritization erosion. Geomorphologically, the *Rarh* laterites (both in situ and ex situ) are now appeared as dissected interfluves having remnant covers of dry deciduous forest. Formerly, the *Rarh Bengal* was the palaeo-valleys of ferruginous depositions which are now appeared as the inverted relief due to duricrust formation and gully erosion.

If we reconstruct the geological clock of laterite evolution, then we can categorized broadly three major events—(1) two widespread volcanisms in Cretaceous Period as Rajmahal Volcanism (~88–118 Ma) and Deccan Volcanism (~69–63 Ma), (2) Palaeocene to Mid-Pliocene in situ intense lateritization and formation of primary high-level laterites (mostly occurred in ~36–26 Ma), and (3) Pliocene to Late Pleistocene ex situ lateritization of deposited ferruginous materials and formation of secondary low-level laterites (up to 35 ka) (Ghosh and Guchhait 2019). Linked with other peninsular laterites, the lateritization age of RBT laterites, gneiss laterites and other primary laterites varies from Palaeocene to Mid-Pliocene on the basis of palaeomagnetism and $^{40}Ar/^{39}Ar$ dating. The region experienced six foremost phases of lateritic weathering and subsequent denudation (<5–25 m Ma^{-1}), viz. ~53–50 Ma,

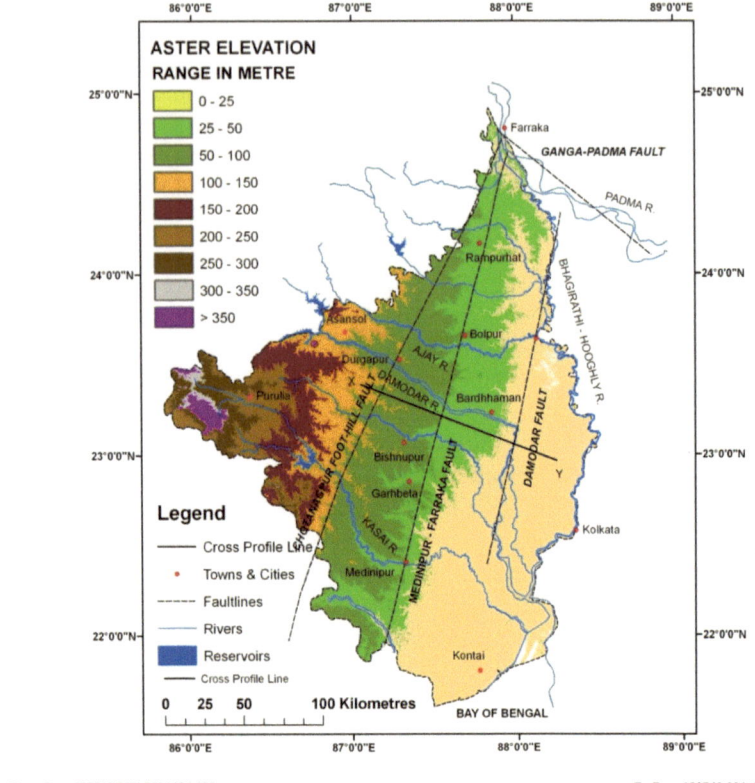

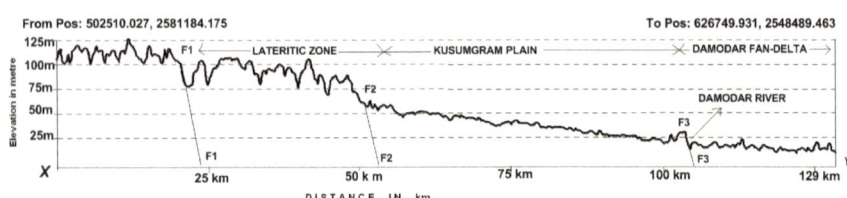

Fig. 8.5 **a** Distribution of laterites (see figure 1) in relation to topography and major basement faults of Shelf zone, viz., GPF (Ganga Padma Fault), CFF (Chotanagpur Foothill Fault), MFF (Medinipur Farakka Fault or Pingla Fault) and DF (Damodar Fault) (Ghosh and Guchhait 2015) in the north-western Bengal Basin (using Landsat ETM+ mosaic SFCC image, 2000–2001), and **b** west to east elevation cross profile (X – Y) with emplacement of faults and development of lateritic *Rarh* region (Ghosh and Guchhait 2015)

~40–32 Ma and ~ 30–23 Ma in the plateau tops and ~47–45 Ma, ~24–19 Ma and ~9 Ma in lowlands, pediments, and valleys (Ghosh and Guchhait 2019). The calculated age of OSL data (ex situ laterites) varies between 150 and 35 ka, thus indicating age of Early Quaternary lateritization (Middle to Late Pleistocene) in western part of the Bengal Basin. In this region, the lateritization processes were favoured by the following factors:

- Onset of tropical wet-dry type of humid climate with more than 1700 mm rainfall (Late Eocene–Early Pleistocene) to accelerate in situ basal weathering, leaching of silica and formation of Fe-mottles;
- Prolonged quiescent phase of geological time with permanent regression of sea level from this region in Early Pleistocene and exposure of valley-filed ferruginous materials;
- Upliftment in between CFF and MFF and its influence on seasonal groundwater regime, sub-aerial exposure, strong leaching, irreversible dehydration, well subsurface drainage, and prolong erosion;
- Adequate permeability and drainage condition to permit deep percolation of silica and deposition of Fe–Al oxides as coating over gravels;
- Warm and strong seasonal climate to hasten the chemical breakdown, hydration, and dehydration of ferruginous and aluminous oxides.

A feedback loop (Fig. 8.6) is inherent in the development of residual and detrital laterites, because lateritization and erosion are continuous cyclic processes since Eocene. The loop involves lowering of primary laterite profiles for accentuation of weathering and consequent accretion of subsequent uplift for facilitating erosion, depositions, re-lateritization, stripping, etchplain formation, and inversion of relief (Ghosh and Guchhait 2019).

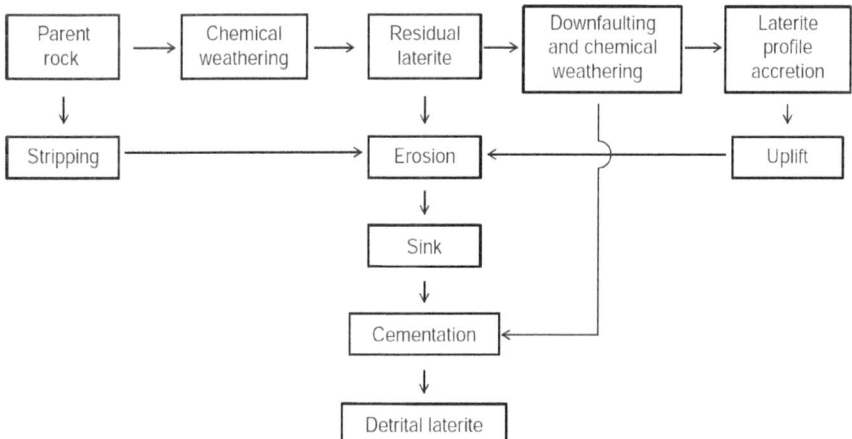

Fig. 8.6 A possible feedback loop to depict sequential development of primary and secondary laterites in the tropical wet-dry climate (Ghosh and Guchhait 2015)

References

Bardossy GY (1981) Paleoenvironments of laterites and lateritic bauxites—effect of global tectonism and bauxite formation. In: Proceedings of the international seminar on lateritisation processes, Trivandum, India

Chatterjee S, Goswami A, Scotese CR (2013) The longest voyage: tectonic, magmatic and palaeoclimatic evolution of the Indian plate during its northward flight from Gondwana to Asia. Gondwana Res 23:238–267

Devaraju TC, Khanadali SD (1993) Laterite bauxite profiles of south western and southern India—characteristics and tectonic significance. Curr Sci 64(11–12):919–921

Ghosh S, Guchhait S (2015) Chraterization and evolution of primary and secondary laterites in northwestern Bengal Basin, West Bengal, India. J Palaeogeogr 4(2):203–230

Ghosh S, Guchhait SK (2019) Modes of formation, Palaeogene to Early Quaternary Palaeogenesis and geochronology of laterites in Rajmahal Basalt Traps and Rarh Bengal of Lower Ganga Basin. In: Das BC, Ghosh S, Islam A (eds) Quaternary geomorphology in India. Springer, Singapore, pp 25–60

Kumar A (1986) Palaeolatitudes and the age of Indian laterites. Palaeogeogr Palaeoclimatol Palaeoecol 53:231–237

Mahapatra S, Dana RK (2009) Lateral variation in gravelly sediments and processes in an alluvial fan—fan-delta setting, north of Durgapur. J Geol Soc India 74(4):480–486

McFarlane MJ (1976) Laterite and landscape. Academic Press, London

Mukherjee B, Rao MG, Karunakaran C (1969) Genesis of kaolin deposits of Birbhum, West Bengal, India. Clay Miner 8:161–170

Niyogi D, Mallick S, Sarkar SK (1970) A preliminary study of laterites of West Bengal, India. In: Chatterjee SP, Das Gupta SP (eds) Selected papers physical geography (vol 1). 21st international geographical congress, Calcutta, National Committee for Geography, pp 443–449

Ollier CD (1991) Laterite profiles, ferricrete and landscape evolution. Zeitschriftfür Geomorphologie 35(2):165–173

Ollier CD (1995) New concepts of laterite formation. Memoirs Geological Society of India, No 32, pp 309–323

Ollier CD, Sheth HC (2008) The high Deccan duricrusts of Indian and their significance for the 'laterite' issue. J Earth Syst Sci 117(5):537–551

Pain CF, Ollier CD (1995) Inversion of relief—a component of landscape evolution. Geomorphology 12:151–165

Sankaran AV, Nambi KSV, Sunta CM (1985) Thermoluminescence of laterites: applicability in dating. Nucl Tracks 17(5):177–183

Schmidt PW, Prasad V, Raman PK (1983) Magnetic ages of some Indian laterites. Palaeogeogr Palaeoclimatol Palaeoecol 44:185–202

Singh LP, Parkash B, Singhvi AK (1998) Evolution of the Lower Gangetic Plain landforms and soils in West Bengal, India. CATENA 33:75–104

Sinha R, Sarkar S (2009) Climate-induced variability in the Late Pleistocene—Holocene fluvial and fluvio-deltaic successions in the Ganga Plains, India. Geomorphology 113(3–4):173–188

Sychanthavong SPH, Patel PK (1987) Laterites and lignites of northwestern India and their relevance to the drift tectonics of the Indian Plate. Curr Sci 56(10):469–473

Tardy Y (1992) Diversity and terminology of laterite profile. In: Martini IP, Chesworth W (eds) Weathering, soils and paleosols. Elsevier, Amsterdam, pp 379–405

Tardy Y, Kobilsex B, Paquet H (1991) Mineralogical composition of geographical distribution of African and Brazilian peri-Atlantic laterites: the influence of continental drift and tropical paleoclimates during the past 150 million years and implications for India and Australia. J Afr Earth Sci 12(1–2):283–295

Vaidyanadhan R, Ghosh RN (1993) Quaternary of the east coast of India. Curr Sci 31(6):231–232

Chapter 9
Laterite and Society

Abstract The study area is presently characterized as low-relief plateau fringe region where the primary and secondary laterites are overlain on the Rajmahal Basalts and towards east this lateritic *Rarh* plain merges with the GBM Delta. Now it is necessary to examine the importance of laterites in the human society and impact of ferruginous properties in resource utilization. In this section, the economic significance of laterites, productivity of latosols, potentiality of geotourism, and soil erosion issue are discussed to get few inferences on the significance of laterite to human.

Keywords Kaolinite · Fire clay · Geotourism · Geosites · Soil erosion

9.1 Economic Significance

The laterite deposits may be described on the basis of the dominant extractable minerals in it: (i) aluminous laterite (bauxite), (ii) ferruginous laterite (iron ore), (iii) manganiferous laterite (manganese ore), (iv) nickeliferous laterite (nickel ore), and (v) chromoferous laterite (chrome ore). It is studied that laterite with Fe_2O_3 : Al_2O_3 ratio more than one and SiO_2 : Fe_2O_3 ratio less than 1.33 is termed as ferruginous laterite, while that having Fe_2O_3 : Al_2O_3 ratio less than one and SiO_2 : Al_2O_3 ratio less than 1.33 is termed as aluminous laterite. The total reserve of laterite deposits are estimated about 559 million tonnes in India. The current reserves of Andhra Pradesh, Gujarat, and Kerala are 4,349,000 tonnes, 9,406,000 tonnes, and 180,000 tonnes, respectively. The profiles of *Rarh* laterite do not have any glimpse of aluminous laterite but these exhibit good deposits of china clay (kaolinite) and fire clay (hydrous silicates of aluminium) at many sites of the Bengal Basin.

Kaolinite (China Clay) is a clay mineral, part of the group of industrial minerals, with the chemical composition $Al_2Si_2O_5(OH)_4$. The main use of 'china clay' is the production of paper—its use ensured the gloss on some grades of coated paper. This clay is also used in ceramics, cosmetics, paints, rubber vulcanization, organic farming, whitewash, and adsorbents in water and wastewater treatment. Fire clay is a mineral aggregate composed of hydrous silicates of aluminium ($Al_2O_3 \cdot 2SiO_2 \cdot 2H_2O$) with or without free silica. It is found that fire clay is resistant to high temperatures, having fusion points higher than 1600 °C. Therefore, it is suitable for lining furnaces, as fire brick, and for manufacture of utensils used in the metal working industries. Because of its ability during firing in the kiln, it can be used to make complex items of pottery such as pipes and sanitary wares.

Laterites and other ferrugnized deposits are considered as essentially of concretionary structure; ferric gels surround the siliceous particles and the hydroxides of iron are absorbed by the clay mineral (mainly kaolinite) particles, sometimes completely covering them. When they are thoroughly soaked with water, saturation is rarely attained due to the several layers of hydrated hydroxide (Gidigasu 1976). So, there are some geo-technical properties of laterites which can be applied to any engineering structures. Laterites are used as building materials, road development, and bank protection. The *murram* deposits are mostly used to develop the un-metal roads of the Indian villages. To protect the coasts and river banks from erosion, the boulders of ferricrete are widely used as barrier and it is found to be sustainable. Laterite soil or latosol acts as natural filter against the trace elements by absorbing it. The researchers of Indian Institute of Technology (Kharagpur, West Bengal) have found that using a chemically treated version of laterite soil, they can enhance its arsenic absorption quality. The laterite-based filter can treat 100 L of arsenic contaminated water per day for five years. It costs only Rs. 4,000/- using minimal maintenance and operation. So, the arsenic-based water filter can be very useful in the arsenic affected areas of GBM delta to provide safe drinking water to each inhabitant.

9.2 Geotourism

Nature is endowed with unique geological features, significant in tracing the expression of Earth's history through ages. The chronicles of the Earth (4.5 billion years old) are inscribed in these unusual physical formations providing an insight into earth science. The richness of various rocks, geological forms, structural events, and landforms has lead to occurrence of diverse geosites offering both scientific and aesthetic interest. So, geotourism is introduced as a part of tourist's activity in which they have the geological patrimony as their main attraction. Their objective is to search for the protected patrimony through the conservation of their resources and of the tourist's environmental awareness. In most of the world, geotourism is purely geologically and geomorphologically focused sustainable tourism as abiotic nature based tourism. Geodiversity is entitled to as geoheritage which refers to the geological features which are inherently or culturally significant offering insight to earth's evolution or history to earth science or that can be utilized for education (INTACH 2016).

Indian National Trust for Art and Cultural Heritage (INTACH 2016) published a monograph on the national geoheritage monuments of India to identify more than 26 geosites for conservation. Geoheritage encompasses global, national, state-wide, and local features of geology at all scales that are intrinsically important sites or culturally important sites offering information or insights into the evolution of the

Earth, or into the history of science, or that can be used for research, teaching, or reference (Joyce 1994; Sharples 1995; Bushy et al. 2001). Kale (2014) has published an edited book, *entitled 'Landscapes and Landforms of India'*, to recognize the unexplored geomorphosites of India which have immense importance to understand the evolutionary history of dynamic earth. Laterite of Angadipuram (Malapuram, Kerala) is also considered as geoheritage site. The National Geological Monument plaque is kept in the PWD (Public Work Department) on a hillock on which the granulite rocks are exposed. These laterites are derived from acid charnockite rocks, and these have economics significance due to mineral deposits of aluminium, iron, and nickel.

Similarly, the laterite-capped Panchgani (19 km east of Mahabaleswar, Maharastra) tableland of the Deccan Traps is now recognized as geomorphosite to observed the breaching and fragmentation of an extensive lateritized surface and subsequent back wearing of the cliffs (Kale 2014). The Deccan volcanic activity ceased about 65 million years ago. The age of Panchgani laterite is Late Cretaceous—Early Paleogene. Therefore, it is possible to identify the laterite deposits as the similar sites of geological and geomorphic importance in the western part of Bengal Basin. The laterites and badlands of Nalhati, Rampurhat, Bolpur, Panagarh, Illambazar, and Gangani are considered as important storehouse of ichnofossils and petrified woods (Figs. 9.1, 9.2 and 9.3). Anyone from earth science can consider these geosites as the natural laboratory to observe in situ or ex situ tropical palaeoweathering profiles and intensive fluvial erosion of badlands.

To make or develop these landscapes as geoheritage or geomorphosite, thirteen principles of National Geographic Society should be followed: (1) integrity of place (reflection of natural heritage), (2) international codes (by the International Council on Monuments and Sites), (3) market selectivity, (4) market diversity, (5) tourist satisfaction, (6) community involvement, (7) community benefit, (8) protection and enhancement of destination appeal, (9) land use, (10) conservation of resources, (11) planning (strive to diversity of the economy and limit population influx), (12) interactive interpretation, and (13) evaluation (by all stakeholders interact and public).

Fig. 9.1 Spectacular view of gully development on the laterites of Rampurhat, Birbhum

Fig. 9.2 Development of badlands over lateritized gravel and pebble deposits at Deul, Bardhaman

Fig. 9.3 Badlands of Gangani over the multi-level laterites at right bank of Silai River, Garhbeta, West Medinipur

9.3 Soil Productivity

Soil–land evaluation is a process of evaluating the soil-site information with a view of assesses its potential for alternative kinds of use. The region of *Rarh* laterites shows few key pedologic features which have direct impact on the agricultural activity and other land uses. In the laterite soils during dry period (November–May), there is a shortage of available water content which is the amount of moisture that can be easily absorbed by plant from a moist soil for its optimum vegetative growth. Due to shallowness, high porosity, low clay content (10–15%), and gravelly character, the lateritic soils have very low to low available water capacity. The soils of *Rarh* are excessively drained because latosol has many pore spaces between particles and high runoff during monsoon. Soil pH varies from 5.8 to 6.2 and it renders plant growth through low cation exchange capacity. The soils have medium–low available organic carbon (0.41–0.80%). It is observed that land capability units have many problems in the soils—(i) low available water capacity, (ii) moderate–high erosion rate, (iii) shallow soil depth, (iv) prolong dryness and surface iron oxide crusting, (v) gravel in soil, and (vi) shallow rooting depth. The prime recommendations to conserve the lateritic soils are as follows: (i) maintain existing plantation, (ii) plantation of low water requirement and quick growth suitable tree species, (iii) grazing control, and (iv) erosion control measures.

In many parts, it is necessary to adopt rainfed agriculture. It is true that most of the past gain in agricultural production have occurred in irrigated conditions. Now it is imperative for us to put more effort for enhancing the agricultural production keeping in view the whopping increase in human population. The present agro-ecological condition, uncertain rainfall, and low productivity compel to install rainfed agriculture. The length of growing periods (moisture condition) varies from 150 to 180 days and the region experiences moisture shortage. In *kharif* season (rainy period), maize, sesame, sunflower, and sugarcane can be grown in the latosols. In many pockets of Birbhum and West Medinipur, the horticulture, like mango plantation, is turned now as popular economic activity.

9.4 Soil Erosion

The laterites of the Bengal Basin are evolved as interfluves in between the peninsular rivers, and the topsoils, as well as subsurface parts, are severely eroded through dense network of rills and gullies. In most cases, secondary ferruginous hardcrusts are appeared as resistant caps over the erodible fluvial sediments, and when the cap is removed by gully erosion, the rest of the underlying sediments is also eroded through defined channels. During the monsoon season and thunderstorms, the rainsplash erosion breaks the aggregate stability of bare topsoils and transports the detached particles downslope through overland flow. It is observed that short period of intense rainstorms are more responsible for vigorous soil erosion than prolonged rainfall.

Table 9.1 Area under different soil classes due to water erosion (>10 tonnes ha^{-1} year^{-1}) in Jharkhand and West Bengal

State	Total area (km^2)	Moderate (10–15) (tonnes ha^{-1} year^{-1}) area (km^2)	Mod. severe (15–20) (tonnes ha^{-1} year^{-1}) area (km^2)	Severe (20–40) (tonnes ha^{-1} year^{-1}) area (km^2)	Very severe (40–80) (tonnes ha^{-1} year^{-1}) area (km^2)	Extreme severe (>80) (tonnes ha^{-1} year^{-1}) area (km^2)	Total area for different classes (km^2)
Jharkhand	79,714	12,424	9,140	16,739	9,748	6,699	51,750
West Bengal	88,752	10,553	3,763	3,257	346	0	17,919

Once the top ferruginous crust or dismantle ferruginous nodules are removed, a network of rills and gullies are generated in the hillslope. In the convex part of lateritic slope, the rills are more dominated, and in concave part or at the end of convex part, the gullies are dominated. The research shows that in Jharkhand the area affected due to water erosion (10 to >80 tonnes ha^{-1} year^{-1}) is 51,750 km^2 and in West Bengal the value is 17,919 km^2 (Table 9.1). The degraded and wastelands of Dumka under different classes are nearly 356,000 ha, and in Birbhum, it is nearly 55,000 ha. Taking steps to prevent or control gully erosion should require no justification, but before taking any action or plan for soil conservation, it is utmost necessary to understand the individuality of rill and gully erosion in the lateritic region and to examine the erosion risk as observed now.

Detachment of soil particles from aggregates primarily by raindrops and flowing water and their transport by runoff water are involved in soil erosion by water. Soil erosion continually shapes and reshapes the land. This is natural or geological erosion, but in Anthropocene, various human actions such as deforestation, overgrazing, unscientific land use, over tilling, and shifting cultivation have accelerated soil erosion beyond the tolerance limit which can vary from 2 to 11 t ha^{-1} year^{-1}. The average maximum potential annual rate of soil loss is 16.25–20.93 kg m^{-2} year^{-1} in the study area. The main erosion processes are rainsplash erosion, inter-rill and rill erosion, gully erosion, and tunnel erosion, but human activities intensify these processes and the erosion risk and vulnerability of land degradation are increased. The erosion has on-site and off-site effects at local scale and regional scale. The on-site effects include loss of topsoil, expansion of bareness, loss of organic matter and nutrients, soil degradation, exposure of plant root, damage of growing crops, etc. The off-site effects are siltation of river bed and reservoirs, eutrophication of ponds and lakes expansion of badlands, etc. (Fig. 9.4).

Based on the empirical study, it is clear that the exposures of secondary laterites are very much prone to erosion, and the major problem is linear erosion through dense network of rills and gullies. The sediment delivery through rills and gullies occurs, if proper erosion and sediment control practices are not installed and maintained.

Fig. 9.4 **a** Reduction of valley incision and stabilization of gully headcut at Baramasia, Rampurhat, and **b** re-vegetation induces stabilization of gully floor and enhances sediment deposition at Maluti, Shikaripara

Fig. 9.5 Soil erosion survey and erosion control measures: **a** field survey during the preparation of dam site using Leica Sprinter 150 m and staff, **b** construction of check dam 5 on January 2017, **c** initial condition of check dam site 2 and 3 on January 2016 and January 2017

The principles provide the ultimate basis for sediment control practices and guiding their selection by—(1) maintaining vegetation cover over barren land, (2) keeping the ground cover by leaving last year's crop residue on the surface, (3) protection against rainfall erosivity and runoff erosivity, (4) incorporating biomass and organic matter into the soil, (5) increasing surface roughness, (6) incorporating bio-engineering measures for erosion management and (7) scientific modification of slope to check flow convergence (Fig. 9.5).

References

Busby AB, Conrads R, Wills P, Roots D (2001) An Australian geographic guide to fossils and rocks. Australian Geographic NSW, Sydney

Gidigasu MD (1976) Laterite soil engineering: pedogenesis and engineering principles. Elsevier, Amsterdam

INTACH (2016) A monograph on National Geoheritage Monuments of India. Indian National Trust for Art and Cultural Heritage, Ministry of Culture, New Delhi

Joyce EB (1994) Geological Heritage Committee. In: Cooper BJ, Branagan DF (eds) Rock me hard—rock me soft—a history of the Geological Society of Australia. Geological Society of Australia, Sydney, pp 30–36

Kale VS (2014) The laterite-capped Panchgani Tableland, Deccan Traps. In: Kale VS (ed) Landscapes and landforms in India. Springer, New York, pp 217–222

Sharples C (1995) Geoconservation in forest management—principles and procedures. Tasforests 7:37–50

Chapter 10
Conclusion

Based on the field and laboratory studies of the important profile sections of western districts (viz., Birbhum, Paschim Bardhaman, Bankura, and West Medinipur) of West Bengal, the laterites are classified into two broad categories—(1) primary or in situ laterites and (2) secondary or ex situ laterites. The primary laterites are genetically related to five types of parent rocks—(1) basalts of Rajmahal Traps, (2) sandstones of Gondwana sequence, (3) gneiss, (4) dolerite, and (5) Tertiary gravels. The ideal weathering profile of primary laterites is found mostly in the province of Rajmahal Basalt Traps (north-eastern part of *Rarh Bengal*) where the ferricrete is pisolitic, massive, blocky, and preserved the columnar structures of basalts. Other ferricretes are generally vermicular and pisolitic type with glimpses of fluid passage like tubes. On the other side, ex situ or transported secondary laterites were developed in the eastern part of *Rarh* Plain, covering mainly in the surroundings of Rampurhat, Bolpur, Labhpur, Guskara, Kanksa, Panagarh, Sonamukhi, Patrasayer, Garhbeta, Lalgarh, Kharagpur, and Rangamati. These laterites are characterized by channel fill deposits and petrified woods with ferruginous materials which are fluvially transported from the eroded province of high-level primary laterites. The relateritization or re-cementation of these materials with gravels, pebbles, and ferricrete nodules developed duricrusts which are geologically comparable with Early–Late Pleistocene Lalgarh Formation (West Medinipur), Kharagpur Formation (East Medinipur), Saltora Formation (Bankura), Worgram Formation (Bardhaman), and Illambazar Formation (Birbhum).

It is finally understood that three basic conditions must be met before enough iron oxide accumulates or segregates to form crust—(1) adequate supply of iron, (2) alternating wet and dry seasons, and (3) iron segregation and accumulation for appreciable periods. It is suggested that the development of both pallid zone and mottled zone is one integrated process and the formation of ferricrete or crust is the final result of that integrated process. If we accept the residuum theory of laterite formation, the original iron precipitates are believed to have formed in the narrow fluctuating range of a groundwater table, which declines as the land surface is lowered. With the cessation of downwasting and stabilization of the groundwater table, the ferruginous residuum is thought to have been hydrated to form massive laterite.

© The Author(s), under exclusive license to Springer Nature Switzerland AG 2020   123
S. Ghosh and S. K. Guchhait, *Laterites of the Bengal Basin*,
SpringerBriefs in Geography, https://doi.org/10.1007/978-3-030-22937-5_10

The mottled zone is formed with the absolute accumulation of iron in kaolinized matrix involving the epigenetic replacement of kaolinite by haematite. Soft nodular and hard nodular iron crusts are described in the upper part of profile that involves the transformation of soft yellow plasma into pisolites. As surface weathering and erosion proceeded, the iron segregations, largely as hematitic mottles with goethite rinds, are progressively exposed at the surface, where they are hardened.

The primary laterites were formed due to in situ lateritization of gneissic rocks, dolerite, Gondwana sediments, Rajmahal Basalt Traps and the vast laterite spread over Barddhmana, Bankura and West Medinipur districts formed in situ by subaerial weathering and alteration of a group of unconsolidated Tertiary sediments. The secondary laterites of *Rarh* Plain were formed due to reworking of earlier formed laterites or iron enrichment of sediments along some palaeo river system of the western Bengal Basin. These laterites are composed of lenticular channel fill deposits, gravels, Fe-pisoids, and larger outsized petrified wood fragments. These lateritized sediments are referred to Pleistocene age (~80–40 ka) of formation. The geochronology of laterites reveals that primary lateritization was started since Eocene–Pliocene and ended at Early Pleistocene, and re-lateritization of ferruginous sediments was restricted up to Middle Pleistocene. Up to Early Miocene, the deep profiles of laterite regolith were widespread in peninsular India, signifying a strong wet–dry tropical palaeoclimatic condition. After that phase in between Late Miocene and Early Pleistocene, the climate was changed to sub-humid to more wet condition. In that event, the deposition of anomalous fluvial gravels–pebbles and lateritization of gravel facies was continued. These ferruginous facies have re-confirmed the prevalence of lateritization climate in the *Rarh Bengal*, related to short-term climate change up to Late Pleistocene. The Late Pleistocene occurrences of colluvio–alluvial calcretized sediments (Older Alluvium—Sijua Formation in West Bengal) and milliolite formation are carrying evidences of arid to semi-arid climatic phases in the Bengal Basin.

Only the profiles of laterite on Upper Jurassic–Early Cretaceous Rajmahal Basal Traps have massive appearance (in situ weathering) reflecting vermicular lateritic crust (i.e. primary laterite), mottled zone with lithomarge clay and deeply weathered basalts. It is suggested that these primary laterites were started to develop in between Late Cretaceous and Eocene. But if we go southward and eastward direction, a ploy-profile of laterites with an intervening erosional surface and original gravels (with iron staining) is present, gradually merging with mottled clay zone and Sijua Formation (Late Pleistocene to Early Holocene). These laterites are identified as the secondary laterites (products of ex situ weathering) which were derived from the primary laterites of western Chota Nagpur Plateau, Raniganj Coal-field, and Rajmahal Basalt Traps towards the Bengal Basin. Due to exposure, well drainage condition and prolong monsoonal wet–dry episode the indurated laterite or consolidated gravely lateritic mass or vesicular or pisolitic laterites were formed at the surface in different parts of northwest Bengal Basin.

Geomorphologically these laterites are found here in two settings—(1) as continuous crusts on plateau fringe where they act as hard caps on the table-like landforms and source of ferralitic materials and (2) at footslopes and interfluves as deposited ferruginous crust, in seepage areas where reduced iron in soil solutions encounters oxidizing conditions and precipitates. The most noticeable feature of south *Rarh* Plain is the presence of a ploy-profile or multi-level lateritic hard crust. It is assumed that bottom profile formed earlier from Upper Tertiary gravelly sediments and after subsequent regional uplift and subsequent erosion developed upper profile of ferruginous alluvium deposits which were lateritized in Late Pleistocene. The age of whole Indian Tertiary laterite decreases from north to south in concordance with the drift tectonic history of the Indian Plate across the equator. So the establishment of strong seasonal monsoon climate (since Eocene), due to equatorward drift and evolution of lofty Himalayas, created the favourable conditions for the development of plateau laterites (Eocene–Miocene) which were subsequently eroded, deposited, and re-lateritized in the valleys within the span of that seasonal climate (Pliocene–Early Pleistocene). From the above perspective of tectono-climatic evolution of ferruginous materials, basically the ex situ type of secondary *Rarh* laterites implies the following palaeogeographic phenomena regarding the genesis:

i. Sea-level fluctuations in the glacial–interglacial epochs,

ii. Re-activation of basements faults due to isostatic disequilibrium of the Bengal Basin (i.e. active passive continental margin of Indian Plate),

iii. Early–Late Quaternary neo-tectonic uplifts, marine regression, and aerial exposure of ferruginous sediments along the shelf zone of Bengal Basin,

iv. Groundwater fluctuation and leaching of silica under the prevalence of tropical wet–dry palaeoclimate, typical to lateritization,

v. Episodic activation of fluvial erosion under more humid climate on the zone primary laterites and fan-deltaic depositions,

vi. Reversal of tropical weathering engine of lixiviation on the exposed detrital ferruginous sediments, and

vii. Re-lateritization of ferruginous minerals as hard crust through formation of polyphased nodules up to Late Pleistocene.

Bibliography

Bland W, Rolls D (1998) Weathering: an introduction to the scientific principles. Arnold, London

Deo SG, Rajaguru SN (2014) Early Pleistocene environment of Acheulian sites in Deccan upland: a geomorphic approach. In: Paddayya K, Deo S G (eds) Recent advances in Acheulian culture studies in India—ISPQS Monograph 6. Indian Society for Prehistoric and Quaternary Studies, Pune, pp 1–22

Eggleton RA (2001) The regolith glossary. CRC LEME, Perth

Goswami AB (1981) Hydrogeology of the lateritic terrain of Bankura and Midnapore districts, West Bengal. In: Proceedings of the international seminar on lateritisation processes, Trivandum, India

Medlicott HB, Blanford WT (1869) A manual of the geology of India, vol 1. Government Press, Calcutta

Miall A (2014) Fluvial depositional systems. Springer, Dordrecht

Ollier CD, Pain C (1996) Regolith, soils and landforms. Wiley, New York

Rudra K (2014) Changing river courses in the western part of the Ganga–Brahmaputra Delta. Geomorphology 15:87–100

Sarkar PR (2004) Rarh—the cradle of civilization. Ananda Nagr Publication, Kolkata

Tardy Y (1993) Petrologie des laterites et des sols tropicaux. Masson, Paris

Widdowson M, Gunnell Y (1999) Lateritization, geomorphology and geodynamics of a passive continental margin: the Konkan and Kanara coastal lowlands of western peninsular India. Spec Publ Int Assoc Sedimentol 27:245–274

Wilson CA, Goodbred SL (2015) Construction and maintenance of the Ganges–Brahmaputra–Meghna Delta: linking process, morphology and stratigraphy. Annu Rev Mar Sci. https://doi.org/10.1146/annurev-marine-010213-135032

Worths H (1905) Weathered dolerite of Rowley Regis (South Staffordshire) compared with the laterite of the Western Ghats near Bombay. Geol Mag 5(2):135–143

© The Author(s), under exclusive license to Springer Nature Switzerland AG 2020
S. Ghosh and S. K. Guchhait, *Laterites of the Bengal Basin*,
SpringerBriefs in Geography, https://doi.org/10.1007/978-3-030-22937-5

Index

© The Author(s), under exclusive license to Springer Nature Switzerland AG 2020   129
S. Ghosh and S. K. Guchhait, *Laterites of the Bengal Basin*,
SpringerBriefs in Geography, https://doi.org/10.1007/978-3-030-22937-5